ASE Test Preparation

Automobile Certification Series

Heating and Air Conditioning (A7)

5th Edition

Australia • Brazil • Japan • Korea • Mexico • Singapore • Spain • United Kingdom • United States

ASE Test Preparation: Automobile Certification Series, Heating and Air Conditioning (A7), 5th Edition

Vice President, Technology and Trades Professional Business Unit: Gregory L. Clayton

Director, Professional Transportation Industry Training Solutions: Kristen L. Davis

Product Manager: Lori Bonesteel

Editorial Assistant: Danielle Filippone

Director of Marketing: Beth A. Lutz

Marketing Manager: Jennifer Barbic

Senior Production Director: Wendy Troeger

Senior Art Director: Benjamin Gleeksman

Content Project Management: PreMediaGlobal

Section Opener Image: Image Copyright Creations, 2012. Used under License from Shutterstock.com

ISBN-13: 978-1-111-12709-1

ISBN-10: 1-111-12709-3

Delmar Cengage Learning
5 Maxwell Drive
Clifton Park, NY 12065-2919
USA

Cengage Learning is a leading provider of customized learning solutions with office locations around the globe, including Singapore, the United Kingdom, Australia, Mexico, Brazil, and Japan. Locate your local office at: **international.cengage.com/region**.

Cengage Learning products are represented in Canada by Nelson Education, Ltd.

For more information on transportation titles available from Delmar, Cengage Learning, please visit our website at **www.trainingbay.cengage.com**.

For more learning solutions, please visit our corporate website at **www.cengage.com**.

Printed at CLDPC, USA, 03-17

Table of Contents

Preface

Delmar, a part of Cengage Learning, is very pleased that you have chosen to use our ASE Test Preparation Guide to help prepare yourself for the Heating and Air Conditioning (A7) ASE certification examination. This guide is designed to help prepare you for your actual exam by providing you with an overview and introduction of the testing process, introducing you to the task list for the Heating and Air Conditioning (A7) certification exam, giving you an understanding of what knowledge and skills you are expected to have in order to successfully perform the duties associated with each task area, and providing you with several preparation exams designed to emulate the live exam content in hopes of assessing your overall exam readiness.

If you have a basic working knowledge of the discipline you are testing for, you will find this book is an excellent guide, helping you understand the "must know" items needed to successfully pass the ASE certification exam. This manual is not a textbook. Its objective is to prepare the individual who has the existing requisite experience and knowledge to attempt the challenge of the ASE certification process. This guide cannot replace the hands-on experience and theoretical knowledge required by ASE to master the vehicle repair technology associated with this exam. If you are unable to understand more than a few of the preparation questions and their corresponding explanations in this book, it could be that you require either more shop-floor experience or further study.

This book begins by providing an overview of, and introduction to, the testing process. This section outlines what we recommend you do to prepare, what to expect on the actual test day, and overall methodologies for your success. This section is followed by a detailed overview of the ASE task list to include explanations of the knowledge and skills you must possess to successfully answer questions related to each particular task. After the task list, we provide six sample preparation exams for you to use as a means of evaluating areas of understanding, as well as areas requiring improvement in order to successfully pass the ASE exam. Delmar is the first and only test preparation organization to provide so many unique preparation exams. We enhanced our guides to include this support as a means of providing you with the best preparation product available. Section 6 of this guide includes the answer keys for each preparation exam, along with the answer explanations for each question. Each answer explanation also contains a reference back to the related task or tasks that it assesses. This will provide you with a quick and easy method for referring back to the task list whenever needed. The last section of this book contains blank answer sheet forms you can use as you attempt each preparation exam, along with a glossary of terms.

OUR COMMITMENT TO EXCELLENCE

Thank you for choosing Delmar, Cengage Learning for your ASE test preparation needs. All of the writers, editors, and Delmar staff have worked very hard to make this test preparation guide second to none. We feel confident that you will find this guide easy to use and extremely beneficial as you prepare for your actual ASE exam.

Delmar, Cengage Learning has sought out the best subject-matter experts in the country to help with the development of *ASE Test Preparation: Automobile Certification Series, Heating and Air Conditioning (A7), 5th Edition.* Preparation questions are authored and then reviewed by a group of certified, subject-matter experts to ensure the highest level of quality and validity to our product.

If you have any questions concerning this guide or any guide in this series, please visit us on the web at **http://www.trainingbay.cengage.com**.

For web-based online test preparation for ASE certifications, please visit us on the web at **http://www.techniciantestprep.com/** to learn more.

ABOUT THE AUTHOR

Jerry Clemons has been around cars, trucks, equipment, and machinery throughout his whole life. Being raised on a large farm in central Kentucky provided him with an opportunity to complete mechanical repair procedures from an early age. Jerry earned an associate in applied science degree in Automotive Technology from Southern Illinois University and a bachelor of science degree in Vocational, Industrial, and Technical Education from Western Kentucky University. Jerry has also completed a master of science degree in Safety, Security, and Emergency Management from Eastern Kentucky University. Jerry has been employed at Elizabethtown Community and Technical College since 1999 and is currently an associate professor for the Automotive and Diesel Technology Programs. Jerry holds the following ASE certifications: Master Medium/Heavy Truck Technician, Master Automotive Technician, Advanced Engine Performance (L1), Truck Equipment Electrical Installation (E2), and Automotive Service Consultant (C1). Jerry is a member of the Mobile Air Conditioning Society (MACS) as well as a member of the North American Council of Automotive Teachers (NACAT). Jerry has been involved in developing transportation material for Delmar, Cengage Learning for seven years.

ABOUT THE SERIES ADVISOR

Mike Swaim has been an Automotive Technology Instructor at North Idaho College, Coeur d'Alene, Idaho since 1978. He is an Automotive Service Excellence (ASE) Certified Master Technician since 1974 and holds a Lifetime Certification from the Mobile Air Conditioning Society. He served as Series Advisor to all nine of the 2011 Automobile/Light Truck Certification Tests (A Series) of Delmar, Cengage Learning ASE Test Preparation titles, and is the author of *ASE Test Preparation: Automobile Certification Series, Undercar Specialist Designation (X1), 5th Edition.*

SECTION

1

The History and Purpose of ASE

ASE began as the National Institute for Automotive Service Excellence (NIASE). It was founded as a nonprofit, independent entity in 1972 by a group of industry leaders with the single goal of providing a means for consumers to distinguish between incompetent and competent technicians. It accomplishes this goal through the testing and certification of repair and service professionals. Though it is still known as the National Institute for Automotive Service Excellence, it is now called "ASE" for short.

Today, ASE offers more than 40 certification exams in automotive, medium/heavy-duty truck, collision repair and refinish, school bus, transit bus, parts specialist, automobile service consultant, and other industry-related areas. At this time there are more than 385,000 professionals nationwide with current ASE certifications. These professionals are employed by new car and truck dealerships, independent repair facilities, fleets, service stations, franchised service facilities, and more.

ASE's certification exams are industry-driven and cover practically every on-highway vehicle service segment. The exams are designed to stress the knowledge of job-related skills. Certification consists of passing at least one exam and documenting two years of relevant work experience. To maintain certification, those with ASE credentials must be re-tested every five years.

While ASE certifications are a targeted means of acknowledging the skills and abilities of an individual technician, ASE also has a program designed to provide recognition for highly qualified repair, support, and parts businesses. The Blue Seal of Excellence Recognition Program allows businesses to showcase their technicians and their commitment to excellence. One of the requirements of becoming Blue Seal recognized is that the facility must have a minimum of 75 percent of its technicians ASE certified. Additional criteria apply, and program details can be found on the ASE website.

ASE recognized that educational programs serving the service and repair industry also needed a way to be recognized as having the faculty, facilities, and equipment to provide a quality education to students wanting to become service professionals. Through the combined efforts of ASE, industry, and education leaders, the nonprofit National Automotive Technicians Education Foundation (NATEF) was created in 1983 to evaluate and recognize academic programs. Today, more than 2,000 educational programs are NATEF certified.

For additional information about ASE, NATEF, or any of their programs, the following contact information can be used:

National Institute for Automotive Service Excellence (NIASE)

101 Blue Seal Drive S.E.

Suite 101

Leesburg, VA 20175

Telephone: 703-669-6600

Fax: 703-669-6123

Website: **www.ase.com**

SECTION

2

Overview and Introduction

Participating in the National Institute for Automotive Service Excellence (NIASE) voluntary certification program provides you with the opportunity to demonstrate you are a qualified and skilled professional technician who has the "know-how" required to successfully work on today's modern vehicles.

EXAM ADMINISTRATION

> *Note:* After November 2011, ASE will no longer offer paper and pencil certification exams. There will be no Winter testing window in 2012, and ASE will offer and support CBT testing exclusively starting in April 2012.

ASE provides computer-based testing (CBT) exams, which are administered at test centers across the nation. It is recommended that you go to the ASE website at ***http://www.ase.com*** and review the conditions and requirements for this type of exam. There is also an exam demonstration page that allows you to personally experience how this type of exam operates before you register.

CBT exams are available four times annually, for two-month windows, with a month of no testing in between each testing window:

- January/February – Winter testing window
- April/May – Spring testing window
- July/August – Summer testing window
- October/November – Fall testing window

Please note, testing windows and timing may change. It is recommended you go to the ASE website at http://www.ase.com and review the latest testing schedules.

UNDERSTANDING TEST QUESTION BASICS

ASE exam questions are written by service industry experts. Each question on an exam is created during an ASE-hosted "item-writing" workshop. During these workshops, expert service representatives from manufacturers (domestic and import), aftermarket parts and equipment manufacturers, working technicians, and technical educators gather to share ideas and convert them into actual exam questions. Each exam question written by these experts must then survive review by all members of the group. The questions are designed to address the practical application of repair and diagnosis knowledge and skills practiced by technicians in their day-to-day work.

After the item-writing workshop, all questions are pretested and quality checked on a national sample of technicians. Those questions that meet ASE standards of quality and accuracy are included in the scored sections of the exams; the "rejects" are sent back to the drawing board or discarded altogether.

Depending on the topic of the certification exam, you will be asked between 40 and 80 multiple-choice questions. You can determine the approximate number of questions you can expect to be asked during the Heating and Air Conditioning (A7) certification exam by reviewing the task list in Section 4 of this book. The five-year recertification exam will cover this same content; however, the number of questions for each content area of the recertification exam will be reduced by approximately one-half.

> *Note:* Exams may contain questions that are included for statistical research purposes only. Your answers to these questions will not affect your score, but since you do not know which ones they are, you should answer all questions in the exam.

Using multiple criteria, including cross sections by age, race, and other background information, ASE is able to guarantee that exam questions do not include bias for or against any particular group. A question that shows bias toward any particular group is discarded.

TEST-TAKING STRATEGIES

Before beginning your exam, quickly look over the exam to determine the total number of questions that you will need to answer. This knowledge will help you gauge your time throughout the exam to ensure you have enough available to answer all of the questions presented. Read through each question completely before marking your answer. Answer the questions in the order they appear on the exam. Leave the questions blank that you are not sure of and move on to the next question. You can return to those unanswered questions after you have finished the others. These questions may actually be easier to answer at a later time, once your mind has had additional time to consider them on a subconscious level. In addition, you might find information in other questions that will help you recall the answers to some of them.

Multiple-choice exams are sometimes challenging because there are often several choices that may seem possible, or partially correct, and therefore it may be difficult to decide on the most appropriate answer choice. The best strategy, in this case, is to first determine the correct answer before looking at the answer options. If you see the answer you decided on, you should still be careful to examine the other answer options to make sure that none seems more correct than yours. If you do not know or are not sure of the answer, read each option very carefully and try to eliminate those options that you know are incorrect. That way, you can often arrive at the correct choice through a process of elimination.

If you have gone through the entire exam and you still do not know the answer to some of the questions, *then guess.* Yes, guess. You then have at least a 25 percent chance of being correct. While your score is based on the number of questions answered correctly, any question left blank or unanswered is automatically scored as incorrect.

There is a lot of "folk" wisdom on the subject of test taking that you may hear about as you prepare for your ASE exam. For example, there are those who would advise you to avoid response options that use certain words such as *all, none, always, never, must,* and *only,* to name a few. This, they claim, is because nothing in life is exclusive. They would advise you to choose response options that use words that allow for some exception, such as *sometimes, frequently, rarely, often, usually, seldom,* and *normally.* They would also advise you to avoid the first and last option (A or D) because exam writers, they feel, are more comfortable if they put the correct answer in the middle (B or C) of the choices. Another recommendation often offered is to select the option that is either shorter or longer than the other three choices because it is more likely to be correct. Some would advise you to never change an answer since your first intuition is usually correct. Another area of folk wisdom focuses specifically on any repetitive patterns created by your question responses (i.e. A, B, C, A, B, C, A, B, C).

Many individuals may say that there are actual grains of truth in this "folk" wisdom, and whereas with some exams this may prove true, it is not relevant in regard to the ASE certification exams. ASE validates all exam questions and test forms through a national sample of technicians, and only those questions and test forms that meet ASE standards of quality and accuracy are included in the scored sections of the exams. Any biased questions or patterns are discarded altogether. Therefore, it is highly unlikely you will actually experience any of this folk wisdom on an actual ASE exam.

PREPARING FOR THE EXAM

Delmar, Cengage Learning wants to make sure we are providing you with the most thorough preparation guide possible. To demonstrate this, we have included hundreds of preparation questions in this guide. These questions are designed to provide as many opportunities as possible to prepare you to successfully attempt and pass your ASE exam. The preparation approach we recommend and outline in this book is designed to help you build confidence in demonstrating what task area content you already know well while also outlining what areas you should review in more detail prior to the actual exam.

We recommend that your first step in the preparation process should be to thoroughly review Section 3 of this book. This section contains a description and explanation of the type of questions you will find on an ASE exam.

Once you understand how the questions will be presented, we recommend that you thoroughly review Section 4 of this book. This section contains information that will help you establish an understanding of what the exam will be evaluating, and specifically, how many questions to expect to be asked in each specific task area.

As your third preparatory step, we recommend you attempt your first preparation exam, located in Section 5 of this book. Answer one question at a time. After you answer each question, review the answer and question explanation information located in Section 6. This section will provide you with instant response feedback, allowing you to gauge your progress, one question at a time, throughout this first preparation exam attempt. If after reading the question explanation you do not feel you understand the reasoning for the correct answer, go back and review the task list overview (Section 4) for the task that is related to that question. Included with each question explanation is a clear identifier of the task area that is being assessed (e.g., Task A.1). If at that point you still do not feel you have a solid understanding of the material, identify a good source of information on the topic, such as an educational course, textbook, or other related source of topical learning, and do some additional studying.

After you have completed your first preparation exam and have reviewed your answers, you are ready to attempt your next preparation exam. A total of six practice exams are available in Section 5 of this book. For your second preparation exam, we recommend that you answer the questions as if you were taking the actual exam. Do not use any reference material or allow any interruptions in order to get a feel for how you will do on the actual exam. Once you have answered all of the questions, grade your results using the answer key in Section 6. For every question that you gave an incorrect answer to, study the explanations to the answers and/or the overview of the related task areas. Try to determine the root cause for your missing the question. The easiest thing to correct is learning the correct technical content. The hardest things to correct are behaviors that lead you to an incorrect conclusion. If you knew the information but still got the question incorrect, there is likely a test-taking behavior that needs to be corrected. An example of this would be reading too quickly and skipping over words that affect your reasoning. If you can identify what you did that caused you to answer the question incorrectly, you can eliminate that cause and improve your score.

Here are some basic guidelines to follow while preparing for the exam:

- Focus your studies on those areas you are weak in.
- Be honest with yourself when determining if you understand something.
- Study often but for short periods of time.
- Remove yourself from all distractions when studying.
- Keep in mind the goal of studying is not just to pass the exam; the real goal is to learn.
- Prepare physically by getting a good night's rest before the exam and eat meals that provide energy but do not cause discomfort.
- Arrive early to the exam site to avoid long waits as test candidates check in.
- Use all of the time available for your exams. If you finish early, spend the remaining time reviewing your answers.
- Do not leave any questions unanswered. If absolutely necessary, guess. All unanswered questions are automatically scored as incorrect.

Here are some items you will need to bring to the exam site:

- A valid, government- or school-issued photo ID
- Your test center admissions ticket
- A watch (not all test sites have clocks)

> *Note:* Books, calculators, and other reference material are not allowed in the exam room. The exceptions to this list are English-Foreign dictionaries or glossaries. All items will be inspected before and after testing.

WHAT TO EXPECT DURING THE EXAM

When taking a CBT exam, as soon as you are seated in the testing center, you will be given a brief tutorial to acquaint you with the computer-delivered test prior to taking your certification exam(s). The CBT exams allow you to only select one answer per question. You can also change your answers as many times as you like. When you select a second answer choice, the CBT will automatically unselect your first answer choice. If you want to skip a question to return to later, you can utilize the "flag" feature, which allows you to quickly identify and review questions whenever you are ready. Prior to completing your exam, you will also be provided with an opportunity to review your answers and address any unanswered questions.

TESTING TIME

Each individual ASE CBT exam has a fixed time limit. Individual exam times will vary based upon exam area and will range anywhere from a half hour to two hours. You will also be given an additional 30 minutes beyond what is allotted to complete your exams to ensure you have adequate time to perform all necessary check-in procedures, complete a brief CBT tutorial, and potentially complete a post-test survey.

You can register for and take multiple CBT exams during one testing appointment. The maximum time allotment for a CBT appointment is four and a half hours. If you happen to register for so many exams that you will require more time than this, your exams will be scheduled into multiple appointments. This could mean that you have testing on both the morning and afternoon of the same day, or they could be scheduled on different days, depending on your personal preference and the test center's schedule.

It is important to understand that if you arrive late for your CBT test appointment, you will not be able to make up any missed time. You will only have the scheduled amount of time remaining in your appointment to complete your exam(s).

Also, while most people finish their CBT exams within the time allowed, others might feel rushed or not be able to finish the test due to the implied stress of a specific, individual time limit allotment. Before you register for the CBT exams, you should review the number of exam questions that will be asked along with the amount of time allotted for that exam to determine whether you feel comfortable with the designated time limitation.

As an overall time management recommendation, you should monitor your progress and set a time limit you will follow with regard to how much time you will spend on each individual exam question. This should be based on the total number of questions you will be answering.

Also, it is very important to note that if for any reason you wish to leave the testing room during an exam, you must first ask permission. If you happen to finish your exam(s) early and wish to leave the testing site before your designated session appointment is completed, you are permitted to do so only during specified dismissal periods.

UNDERSTANDING HOW YOUR EXAM IS SCORED

You can gain a better perspective about the ASE certification exams if you understand how they are scored. ASE exams are scored by an independent organization having no vested interest in ASE or in the automotive industry. With CBT exams, you will receive your exam scores immediately.

Each question carries the same weight as any other question. For example, if there are 50 questions, each is worth 2 percent of the total score.

Your exam results can tell you:

- Where your knowledge equals or exceeds that needed for competent performance, or
- Where you might need more preparation.

Your ASE exam score report is divided into content "task" areas; it will show the number of questions in each content area and how many of your answers were correct. These numbers provide information about your performance in each area of the exam. However, because there may be a different number of questions in each content area of the exam, a high percentage of correct answers in an area with few questions may not offset a low percentage in an area with many questions.

It should be noted that one does not "fail" an ASE exam. The technician who does not pass is simply told "More Preparation Needed." Though large differences in percentages may indicate problem areas, it is important to consider how many questions were asked in each area. Since each exam evaluates all phases of the work involved in a service specialty, you should be prepared in each area. A low score in one area could keep you from passing an entire exam. If you do not pass the exam, you may take it again at any time it is scheduled to be administered.

There is no such thing as average. You cannot determine your overall exam score by adding the percentages given for each task area and dividing by the number of areas. It does not work that way because there generally is not the same number of questions in each task area. A task area with 20 questions, for example, counts more toward your total score than a task area with 10 questions.

Your exam report should give you a good picture of your results and a better understanding of your strengths and areas needing improvement for each task area.

SECTION 3

Types of Questions on an ASE Exam

Understanding not only what content areas will be assessed during your exam, but how you can expect exam questions to be presented, will enable you to gain the confidence you need to successfully pass an ASE certification exam. The following examples will help you recognize the types of question styles used in ASE exams and assist you in avoiding common errors when answering them.

Most initial certification tests are made up of between 40 and 80 multiple-choice questions. The five-year recertification exams will cover the same content as the initial exam; however, the actual number of questions for each content area will be reduced by approximately one-half. Refer to Section 4 of this book for specific details regarding the number of questions to expect to receive during the initial Heating and Air Conditioning (A7) certification exam.

Multiple-choice questions are an efficient way to test knowledge. To correctly answer them, you must consider each answer choice as a possibility, and then choose the answer that *best* addresses the question. To do this, read each word of the question carefully. Do not assume you know what the question is asking until you have finished reading the entire question.

About 10 percent of the questions on an actual ASE exam will reference an illustration. These drawings contain the information needed to correctly answer the question. The illustration should be studied carefully before attempting to answer the question. When the illustration shows a system in detail, look over the system and try to figure out how the system works before you look at the question and the possible answers. This approach will ensure you do not answer the question based upon false assumptions, or partial data, but instead have reviewed the entire scenario being presented.

MULTIPLE-CHOICE/DIRECT QUESTIONS

The most common type of question used on an ASE exam is the direct multiple-choice style question. This type of question contains an introductory statement, called a stem, followed by four options: three incorrect answers, called distracters, and one correct answer, the key.

When the questions are written, the point is to make the distracters plausible to draw an inexperienced technician to inadvertently select one of them. This type of question gives a clear indication of the technician's knowledge.

Here is an example of a direct style question:

1. Which device is used as the input to cause the A/C compressor clutch to disengage during a parallel parking maneuver?

   A. Coolant temperature sensor
   B. Throttle position sensor
   C. Oxygen sensor
   D. Power steering switch

TASK A.24

Answer A is incorrect. The coolant sensor is used to sense engine temperature. Some engine control modules will deactivate the AC compressor clutch if engine temperature rises to near overheating range.

Answer B is incorrect. The throttle position sensor is used to sense throttle angle. Some engine control modules will deactivate the AC compressor clutch if the throttle angle reaches very high levels.

Answer C is incorrect. The oxygen sensor is used to provide feedback of oxygen content in the exhaust.

Answer D is correct. The power steering switch is used to provide power steering information to the engine control modules. Some engine control modules deactivate the AC compressor clutch when this switch senses high pressure; such as when parallel parking.

COMPLETION QUESTIONS

A completion question is similar to the direct question except the statement may be completed by any one of the four options to form a complete sentence. Here is an example of a completion question:

2. The vacuum valve in the radiator cap is stuck closed. The result of this problem could be:

   A. The upper radiator hose collapses after the engine is shut off
   B. Excessive cooling system pressure at normal engine temperature
   C. The engine overheats when operating under a heavy load
   D. The engine overheats during extended idle periods

Answer A is correct. The upper radiator hose would collapse due to a vacuum being present in the radiator. The vacuum valve operates by allowing coolant to return to the radiator after the engine has cooled down.

Answer B is incorrect. A problem with the pressure release valve would not cause the cooling system to develop excessive pressure.

Answer C is incorrect. There are several possible causes for engine overheating, but a vacuum valve on the radiator cap would not cause overheating.

Answer D is incorrect. An inoperative cooling fan is the likely cause of overheating during long idle times.

TECHNICIAN A, TECHNICIAN B QUESTIONS

TASK C.1.5

This type of question is usually associated with an ASE exam. It is, in fact, two true-false statements grouped together, such as: "Technician A says…" and "Technician B says…", followed by "Who is correct?

In this type of question, you must determine whether either, both, or neither of the statements are correct. To answer this type of question correctly, you must carefully read each technician's statement and judge it on its own merit.

Sometimes this type of question begins with a statement about some analysis or repair procedure. This statement provides the setup or background information required to understand the conditions about which Technician A and Technician B are talking, followed by two statements about the cause of the concern, proper inspection, identification, or repair choices. Analyzing this type of question is a little easier than the other types because there are only two ideas to consider, although there are still four choices for an answer.

Again, Technician A, Technician B questions are really double true-or-false questions. The best way to analyze this type of question is to consider each technician's statement separately. Ask yourself, "Is A true or false? Is B true or false?" Once you have completed an individual evaluation of each statement, you will have successfully determined the correct answer choice for the question, "Who is correct?"

An important point to remember is that an ASE Technician A, Technician B question will never have Technician A and B directly disagreeing with each other. That is why you must evaluate each statement independently.

An example of a Technician A/Technician B style question looks like this:

3. An A/C compressor is being diagnosed for a slipping front clutch drive plate. A voltage test at the compressor clutch coil shows 8.5 volts when the compressor is engaged. Technician A says that the A/C clutch relay could have high resistance on the load side contacts. Technician B says that A/C coil ground could be loose. Who is correct?

   A. A only
   B. B only
   C. Both A and B
   D. Neither A nor B

Answer A is incorrect. Technician B is also correct.

Answer B is incorrect. Technician A is also correct.

Answer C is correct. Both Technicians are correct. High resistance on relay contacts could cause the compressor clutch coil to not receive the full system voltage. A loose A/C coil ground could cause the coil to not receive full system voltage. Voltage drop tests could be performed with the system energized to further pinpoint the problem.

Answer D is incorrect. Both Technicians are correct.

EXCEPT QUESTIONS

TASK B.1.4

Another type of question used on ASE exams contains answer choices that are all correct except for one. To help easily identify this type of question whenever it is presented in an exam, the word "EXCEPT" will always be displayed in capital letters. Furthermore, a cautionary statement will alert you to the fact that the next question is different from the ones otherwise found in the exam. With the EXCEPT type of question, only one incorrect choice will actually be listed among the options, and that incorrect choice will be the key to the question. That is, the incorrect statement is counted as the correct answer for that question.

Be careful to read these question types slowly and thoroughly; otherwise, you may overlook what the question is actually asking and answer the question by selecting the first correct statement.

An example of this type of question would appear as follows:

4. Any of these faults in the A/C system could cause an elevated high side reading EXCEPT:

   A. Refrigerant overcharge
   B. Restricted airflow to the condenser
   C. Poor airflow across the evaporator
   D. A slipping fan clutch

Answer A is incorrect. An overcharged A/C system would cause elevated high side pressures.

Answer B is incorrect. A condenser airflow problem would cause elevated high side pressures.

Answer C is correct. Poor airflow across the evaporator core would not cause elevated high side pressure.

Answer D is incorrect. A slipping fan clutch could cause elevated high side pressures when the vehicle is at low speeds or is sitting still.

LEAST LIKELY QUESTIONS

TASKS A.2, B.2.8

LEAST LIKELY questions are similar to EXCEPT questions. Look for the answer choice that would be the LEAST LIKELY cause of the described situation. To help easily identify this type of question, whenever they are presented in an exam, the words "LEAST LIKELY" will always be displayed in capital letters. In addition, you will be alerted before a LEAST LIKELY question is posed. Read the entire question carefully before choosing your answer.

An example of this type of question is shown below:

5. Which of these would be the LEAST LIKELY location to find technical repair literature for a late-model vehicle's air conditioning problem?

   A. Repair manua
   B. Technical CD-ROM
   C. Microfiche retrieval system
   D. Technical repair website

Answer A is incorrect. Service literature is sometimes available in a printed repair manual.

Answer B is incorrect. Service literature is sometimes available on a technical CD ROM.

Answer C is correct. Microfiche retrieval systems are a very old method of data storage and would not be used to find repair information on a late-model vehicle.

Answer D is incorrect. Service literature is sometimes available on a technical repair website.

SUMMARY

The question styles outlined in this section are the only ones you will encounter on any ASE certification exam. ASE does not use any other types of question styles, such as fill-in-the-blank, true/false, word-matching, or essay. ASE also will not require you to draw diagrams or sketches to support any of your answer selections, although any of the described question styles may include illustrations, charts, or schematics to clarify a question. If a formula or chart is required to answer a question, it will be provided for you.

SECTION 4

Task List Overview

INTRODUCTION

This section of the book outlines the content areas or "task list" for this specific certification exam, along with a written overview of the content covered in the exam.

The task list describes the actual knowledge and skills necessary for a technician to successfully perform the work associated with each skill area. This task list is the fundamental guideline you should use to understand what areas you can to expect to be tested on, and how each individual area is weighted to include the approximated number of exam questions you can expect to be given for that area during the ASE certification exam. It is important to note that the number of exam questions for a particular area is truly to be used as a guideline only. ASE advises that the questions on the exam may not equal the number specifically listed on the task list. The task lists are specifically designed to tell you what ASE expects you to know how to do and to help you be ready to be tested.

Similar to the role this task list will play in regard to the actual ASE exam, Delmar developed the six preparation exams, located in Section 5 of this book, using this task list as a guideline as well. It is important to note that although both ASE and Delmar use the same task list as a guideline for creating our test questions, none of the test questions you will use in our practice exams will be found in the actual, live ASE exams. This is a true statement for any test preparatory material you use. ASE reserves the actual exam questions so they are *only* ever visible during actual exams.

Task List at a Glance

The Heating and Air Conditioning (A7) task list focuses on four core areas, and you can expect to be asked a total of approximately 50 questions on your certification exam, broken out as outlined:

A. Heating, Ventilation, A/C (HVAC) and Engine Cooling System Service, Diagnosis, and Repair (21 Questions)

B. Refrigeration System Component Diagnosis and Repair (10 questions)

C. Operating Systems and Related Controls Diagnosis and Repair (19 questions)

Based upon this information, the following graphic is a general guideline demonstrating which areas will have the most focus on the actual certification exam. This data may help you prioritize your time when preparing for the exam.

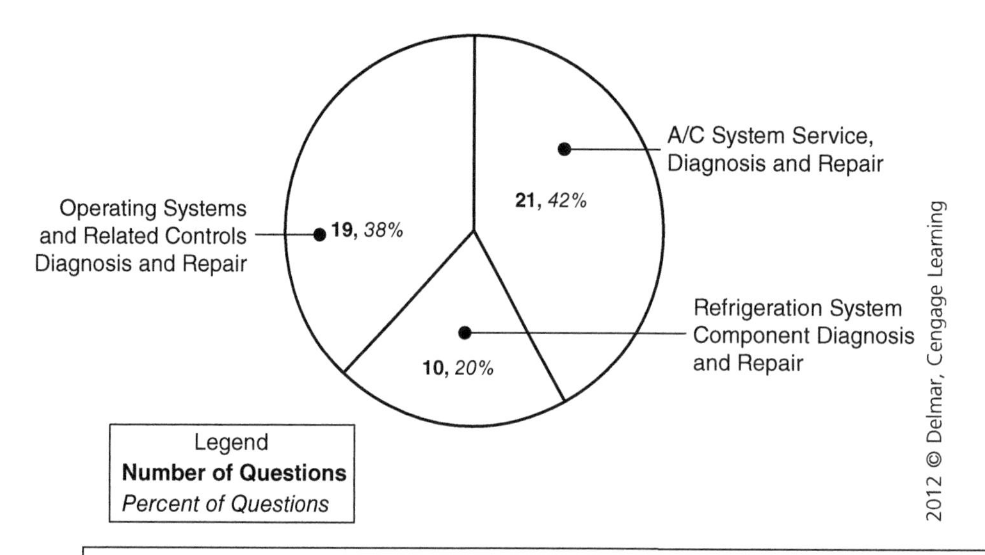

Note: There could be additional questions that are included for statistical research purposes only. Your answers to these questions will not affect your test score, but since you do not know which ones they are, you should answer all questions in the test. The five-year Recertification Test will cover the same content areas as those listed above. However, the number of questions in each content area of the Recertification Test will be reduced by one-half.

A. Heating, Ventilation, A/C (HVAC) and Engine Cooling System Service, Diagnosis, and Repair (21 Questions)

1. Identify system type and conduct performance test on the A/C system; determine needed repairs.

The two common types of A/C systems being used today are the orifice tube system and the thermal expansion valve (TXV) system. The simplest way of differentiating one system from another is by where the drying device is located. For example, if the receiver/drier is in the high side of the system, then the system is a TXV system. Conversely, if the accumulator drier is in the low side of the system, then the system is an orifice tube system.

Conducting performance tests is a common activity for an HVAC technician. The steps to follow include:

- Connect a pressure manifold set to the vehicle high- and low-side test fittings.
- Insert a thermometer into the center A/C duct.
- Adjust the A/C controls to Max A/C (recirculate mode) with the blower on high speed.
- Set the temperature to the coldest setting (both zones should be set to the coldest temperature setting if equipped with dual zone controls).
- Increase and hold the engine speed between 1500 and 2000 rpm.
- Shut all of the windows and doors on the vehicle being tested.
- After five minutes, read the low- and high-side pressures and the duct temperature.

Test results will vary depending on the ambient temperature and humidity level. The duct temperature should be about 40°F to 50°F (4.4°C – 10°C). The system pressures should be approximately 20 to 40 psi on the low side. The high-side pressure will vary greatly depending on the conditions. Generally speaking, the high-side pressure will be about 2 times the ambient temperature with low to moderate humidity levels and as much as 3 times the ambient temperature with higher levels of humidity. For example, normal pressures on a 70°F (21°C) day with 30 percent humidity would be in the range of 30 psi on the low side and 140 psi on the high side. Conversely, if the humidity level were 80 percent on the same 70°F (21°C) day, then the high-side pressures would likely be in the 210 psi range.

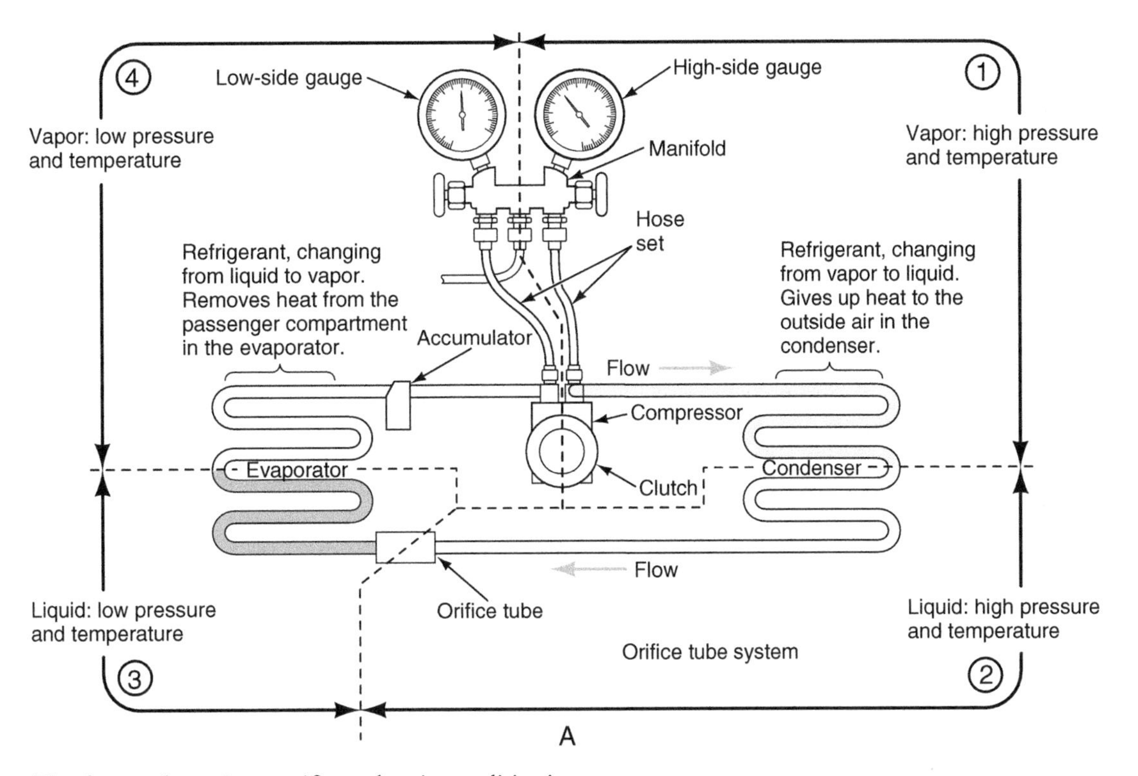

The figure above is an orifice tube air conditioning system.

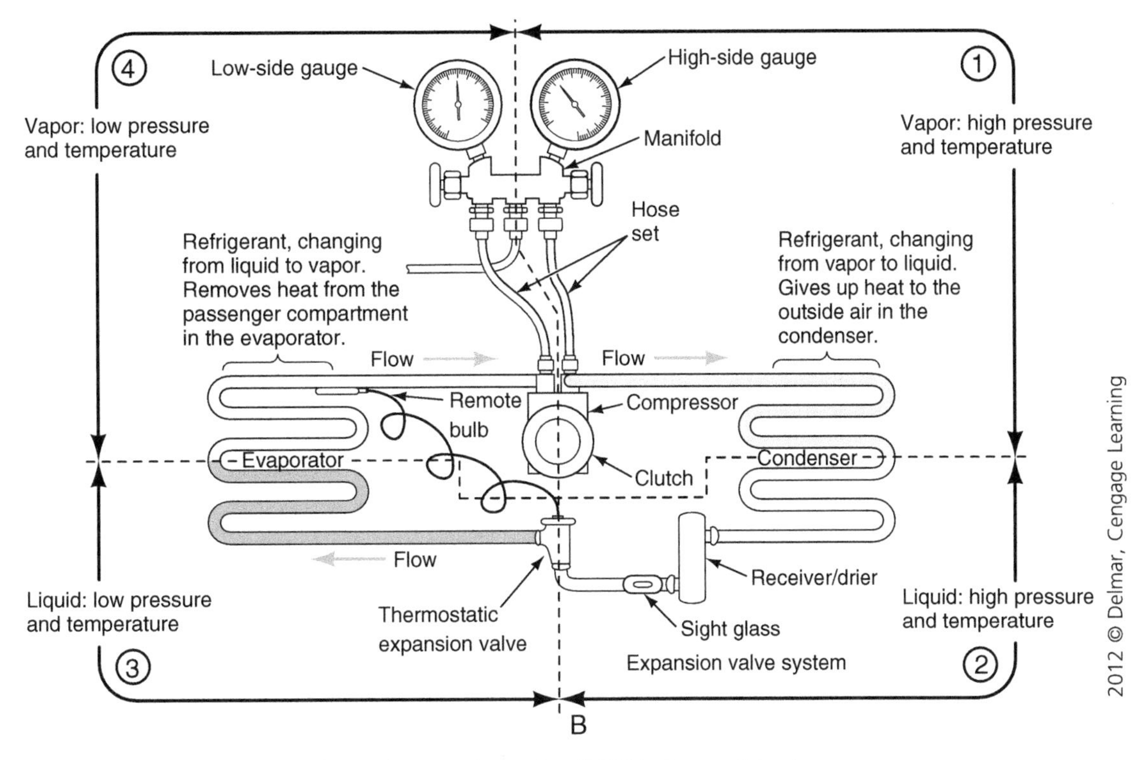

The figure above shows a thermal expansion valve (TXV) air conditioning system.

2. Diagnose A/C system problems indicated by system pressures and/or temperature readings; determine needed repairs.

There are many potential causes for a vehicle to display irregular performance and unusual system pressures. The following list details common system faults accompanied with the typical pressure characteristics.

- Refrigerant overcharge: Slightly elevated low-side pressure with very elevated high-side pressure.
- Airflow problem at the condenser: Slightly elevated low-side pressure with very elevated high-side pressure.
- Refrigerant undercharge: Low pressure on both sides of the system and a short cycling compressor.
- Metering device restriction: Low pressure on both sides of the system and a short cycling compressor.
- Faulty compressor (compressor not engaging): Pressures equal on both sides.
- Faulty compressor (reed valve damage): Elevated low-side pressure and reduced high-side pressure.
- High-side internal restriction: Unusual pressures accompanied by a cold area at some point in the high side (such as in the condenser or at the receiver/drier).
- Excessive air in the A/C system: Slightly elevated pressures on both sides of the system and a reduced cooling performance during high ambient temperatures.
- Variable displacement compressor stuck at high output: Normal high-side pressures but reduced low-side pressures after several minutes of operation. System will likely have ice on the suction line and possibly ice coming from the vents.
- Variable displacement compressor stuck at low output: Low pressures on both sides of the system with warm air coming from the vents.
- Deicing system not working: Normal operation during the first few minutes of running the AC. Then the low side pressures will be low and there likely will be frosting on the suction line as well as the evaporator core freezing up. A frozen evaporator core can produce ice chips from the vents and could lead to loss of airflow at the vents.

3. Diagnose A/C system problems indicated by sight, sound, smell, and touch procedures; determine needed repairs.

There are general characteristics that a normal operating A/C system should possess. These include:

- Components on the low side should be cool or cold, but not covered with ice.
- Components on the high side should be hot or very warm.
- The line entering the condenser should be about 20°F to 50°F hotter than the line exiting the condenser.
- The HVAC drain tube should produce water after several minutes of operation in humid conditions.
- On a cycling A/C system, the compressor will cycle on and off as needed to prevent evaporator freeze-up.

4. Leak test A/C system; determine needed repairs.

Refrigerant leaks account for the largest number of problems that occur in A/C systems. HVAC technicians have to be very good at finding leaks as well as repairing leaks. Methods of finding leaks are listed below:

- Visual inspection: Listen for hissing noises of refrigerant escaping the system. Look for signs of oily residue on any A/C component. Whenever there is evidence of refrigerant oil on items such as lines, hoses, fittings, service valves, compressors, and other components it is very likely that refrigerant is leaking at these locations, too.
- Electronic detector: With at least 70 psi of static pressure in the A/C system, a technician can use the electronic leak detector to find A/C leaks. Turn on the detector and let it warm up. Move the wand slowly throughout the A/C system, being careful to not get the tip dirty. It is advisable to hold the wand on the bottom side of the A/C components due to the weight of the refrigerant. Whenever the detector changes its tone, a refrigerant leak is present and should be repaired. Remember to always check the service fittings during all leak inspections. These fittings often get overlooked in the leak detection process.
- Nitrogen injection: Using an inert pressurized gas is a viable method of leak detection, especially if the system is already empty. It is advisable to inject about 120 psi of nitrogen into the system. After injecting the nitrogen, the technician needs to watch the pressures to see if any pressure loss is occurring. It is also a good idea to use a listening device to check for any hissing sounds coming from the A/C system. Finally, a soap spray solution can be applied to the A/C components to see if any bubbling occurs.
- Dye and black-light: Injecting a fluorescent dye into the A/C system and then operating the system is a very accurate way to find leaks in an A/C system. Some manufacturers are now adding dye into the vehicles on the assembly line. A black-light is needed to view the A/C system in order to amplify the brightness of the dye. If an evaporator core is suspected to be leaking, be sure to shine the black-light on the case drain tube to look for the presence of the dye in the condensate water.
- Vacuum pump: An effective way to find A/C leaks before recharging the system is by using the vacuum pump. After evacuating the system, simply watch the low-side gauge for a few minutes to see if any vacuum is lost. A sealed system should hold a tight vacuum during this test. If vacuum does bleed off, then the system should not need to be recharged. Further leak testing should be performed.

5. Identify A/C system refrigerant and existing charge amount; recover refrigerant.

It is advisable to connect a refrigerant identifier to every system before any repair work is performed to assure that the refrigerant is not mixed with other chemicals. If the refrigerant is acceptable, then the technician can verify the charge amount by recovering with a recovery station that has accurate scales. It is vital to charge late-model A/C systems to the exact charge amount. Most vehicles will have a decal under the hood that has the correct charge amount. A quality A/C charging station will work well to accomplish this task. If a charging station is not available, then an electronic scale will work well to get the system charged to the correct level.

6. Evacuate A/C system.

The A/C system needs to be evacuated for at least 30 minutes any time that the system has been open to the atmosphere. The process of pulling a good vacuum on an A/C system causes any moisture/humidity to change into a vapor and be sucked into the vacuum pump. It is critical to remove all of the moisture from the A/C system to prevent any acidic activity to occur when the system is recharged with refrigerant.

The process of evacuation is also a tool to check for leaks in the A/C system. After the vacuum pump is shut off, the system should be monitored to see if the deep vacuum is held in the system. If the vacuum starts bleeding off, then the system likely has a leak that will need to be located and repaired. Another indication of a potential A/C system leak is if the vacuum pump does not draw a deep vacuum (25–29 in. Hg) within 10 to 15 minutes of operation.

7. Inspect A/C system components for contamination; determine needed repairs.

Before servicing a system that is unknown, the technician should use a refrigerant identification machine to determine the contents of the system before potentially contaminating the A/C recovery machine with blended or contaminated refrigerants. With the significant misinformation that was and still is present in some circles, there are many "retrofitted" systems out there that contain everything from R-12 oils and R-134a refrigerant to pure butane in them. If an A/C system has been contaminated, the technician must decide what action to take. It is not advisable to pull contaminated refrigerant into an A/C recovery machine because it will ruin the pure refrigerant that is in the machine. Some shops have a dedicated A/C recovery machine that is only used on systems with contamination problems.

Technicians must also be aware of another potential problem, the presence of sealer in the A/C system. Most electronic identifiers will not detect these sealers. There is a test kit available to test for these sealers. It is highly advisable to perform this test any time that it is suspected that someone added his/her own refrigerant out of small cans. If sealer is detected in a system, then do not recover the refrigerant into a regular A/C recovery machine. It is recommended to use a recovery machine that has a special filter installed that will remove the sealer additive.

8. Charge A/C system with refrigerant (liquid or vapor).

There are several potential ways of charging refrigerant into an A/C system. The quickest and most accurate way to perform this process is using a machine that combines the recovery, evacuation, and recharging processes into one housing. These machines have built-in scales that are very accurate to assist the technician in charging the exact amount of refrigerant vapor back into the A/C system. In most cases, the technician does not have to operate the A/C system while recharging with these types of machines. It is just necessary to program the evacuation time and then the charge amount; the machine runs the process and completes the recharge. After the refrigerant is installed, it is wise to operate the A/C system to monitor pressures and performance. If the system performs well, then the machine can be disconnected and the caps put back onto the service fittings.

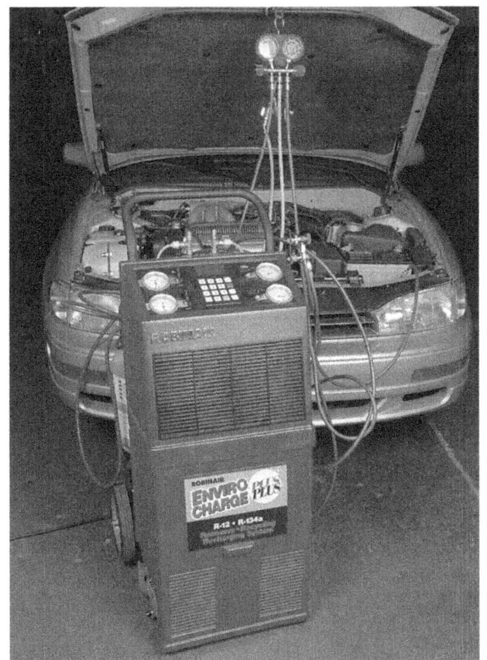

If a multi-function A/C machine is not available, then the technician can still recharge the A/C system using either small cans or the larger cylinders. With either source, the technician will need to pull a vacuum on the system with a vacuum pump connected to the manifold set. After the vacuum is pulled, the yellow service hose is removed from the vacuum pump and connected to the refrigerant source. Refrigerant can be charged into both the low and high sides of the A/C system initially with the system off. After the pressures equalize, then the high-side manifold valve must be closed and the A/C system can be operated. It is important to complete the charging process by charging as a vapor. This means that the refrigerant containers need to be sitting upright and not inverted. If the larger cylinders are used, it is important to use some type of scale to measure how much refrigerant is being charged. The small can method is the least accurate way of recharging an A/C system, due to the fact that there is not a good way to determine how much refrigerant fails to enter the A/C system when the cans are switched out.

9. Identify A/C system lubricant type and capacity; replenish lubricant if necessary.

All (nonhybrid) R-134a A/C systems use some type of polyalkylene glycol (PAG) refrigerant oil. This oil is synthetic and comes in several viscosities, so care should be taken when adding refrigerant oil to an A/C system. The refrigerant decal located under the hood is a good place to find the oil information. If this decal has been removed, then a reliable service information resource can be used to find the recommended oil and the oil capacity. It should be noted that refrigerant oil should be installed whenever A/C components are replaced or after being flushed. The key to remember is to distribute the oil throughout the system, not all in one place.

10. Inspect and replace passenger compartment (cabin air, pollen) filter.

A cabin air filter is used on the inlet of the air distribution system to clean foreign materials from the air before it reaches the evaporator. Two types of filters are common: filter paper and absorption. Both can remove very small particles from the inlet air including pollen, mold spores, and other allergens that can affect passenger health and comfort.

These filters are usually located under the hood at the outside air intake point, under the dash, or in both places. Replacement should be at scheduled maintenance intervals. Driving conditions, geographic location, and system usage will alter the frequency of needed service. Typically the recommended replacement interval is between 15,000 and 24,000 miles (24,140 and 38,624 km). A visual inspection is the standard way to evaluate the need for replacement. If the filter becomes clogged, reduced airflow through the evaporator can affect system performance and possibly result in evaporator freeze-up.

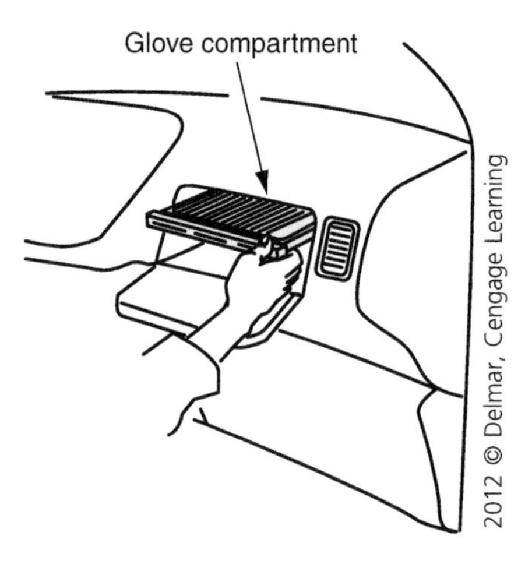

2012 © Delmar, Cengage Learning

11. Disarm and enable the airbag system for vehicle service following manufacturer's recommended procedures.

Frequently an HVAC technician has to work around components of the airbag system. These safety systems can be very dangerous if the proper precautions are not taken. Disabling and enabling the airbag system will vary from vehicle to vehicle. It is always advisable to consult the available service information to get the exact procedures for each vehicle that is repaired.

To disable the airbag system on most vehicles, the negative battery cable will need to be disconnected and secured to prevent accidental reconnection. This can be done by installing black electrical tape around the terminal to prevent it from touching the battery. It is also recommended to wait about 10 minutes after disconnecting the cable before beginning any work. Most airbag control modules contain capacitors that remain powered up for a short period of time after the battery is disconnected.

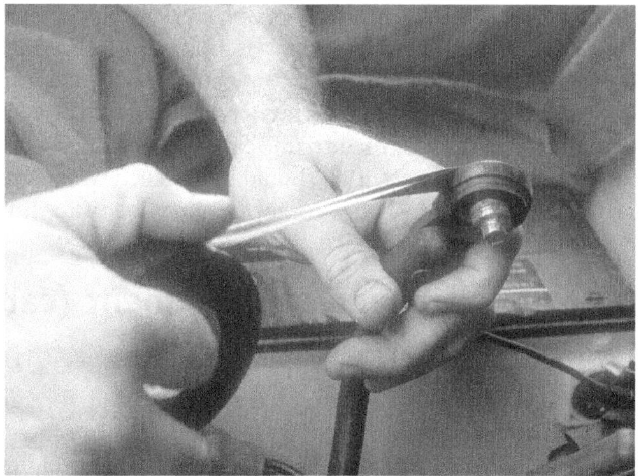

2012 © Delmar, Cengage Learning

Extreme care should be taken if live airbag components have to be handled during an HVAC repair. Always hold the airbag components away from your body and watch your step when walking with them. When storing airbag assemblies, be sure to set them "face up" in a secure location.

After the repair is made and the airbag system needs to be enabled, the technician should maintain a safe distance from the devices while reconnecting them. Once the devices are all connected, it is time to reconnect the battery. Carefully turn on the ignition switch and watch the airbag indicator. The airbag indicator should illuminate for a few seconds and then turn off. If the light stays on, the system will need to be scanned and diagnosed for a fault.

12. Read diagnostic trouble codes (DTCs) and interpret scan tool data stream.

Most late model vehicles are equipped with computers that control the operation of the heating and air conditioning system. These computer systems continually run self-diagnostics and set diagnostic trouble codes (DTCs) when certain problems occur. The technician must be able to retrieve the trouble codes to diagnose the problem. Some systems allow the technician to retrieve DTCs with flash code diagnostics. The technician activates this process by depressing certain buttons on the climate control head. Once activated, these systems will display trouble codes on the electronic display or flash an indicator in a sequence that can be interpreted as a code.

Scan tools can also be used to retrieve DTCs from the climate control system. The scan tool connects to the vehicle at the data link connector (DLC). Scan tools can be either factory-based or generic. Factory-based scan tools will typically only work on a particular vehicle make. Generic scan tools will typically function on a variety of vehicle makes. After the DTC is retrieved, the technician will need to diagnose what caused the DTC to set. Scan tools often display live sensor data for the technician to view and interpret. A good technician will get accustomed to understanding what normal values look like. If a sensor does not produce normal values on the data stream, then the technician will perform further tests at the sensor and associated wiring and connections.

13. Read and interpret technical literature (service publications and information including wiring schematics).

Automotive technicians need to be able to think logically and read technical literature very well. Technical literature can be sometimes found in a book, but most of the technical data that technicians use today is in an electronic format. A successful technician has the ability to use a computer to retrieve repair information, flow charts, technical service bulletins, and wiring schematics. Once the correct piece of repair information is found, it is very important to carefully read and follow the directions found in the repair document.

Typically, wiring schematics use many common characteristics or conventions. See the following list for some helpful hints to follow when viewing wiring schematics:

- Power (B+) comes from the top of the page.
- Ground (negative) comes from the bottom of the page.
- Wire colors are represented by letters next to the wires.
- Circuit numbers are represented by numbers next to the wires.
- Universal symbols for fuses, switches, relays, motors, resistors, and connectors are used to show the circuit layout.

It is a good diagnostic practice to trace the flow of electricity in order to understand how the circuit works and then decide where to start the diagnosis.

14. Use a scan tool, digital multimeter (DMM), or digital storage oscilloscope (DSO) to inspect or test HVAC system sensors, actuators, circuits, and control modules; determine needed repairs.

Scan tools can be used to communicate with the many computers found on late-model vehicles. They can be used to retrieve DTCs, Freeze Frame data, as well as to view live data from sensors, switches and actuators. In addition to monitoring the inputs and codes in a computer system, the scan tool can be used to perform output tests on the motors, actuators, and relays that are controlled by the computer. Performing output tests with a scan tool is a very good method to locate the cause of some electrical faults that can occur in the HVAC system. Using a scan tool is a very efficient method of troubleshooting the HVAC system.

It is often necessary to use a digital multimeter (DMM) to more closely test the circuits and components of the HVAC system. The DMM can be used to measure voltage, amperage, and resistance in the circuit in question. Many times the technician will need to consult some type of service literature to perform accurate and thorough tests on the HVAC system. See Task A.13 for more information on service literature.

A digital storage oscilloscope (DSO) is occasionally used to closely monitor electrical activity in HVAC electrical and electronic circuits. A DSO can be useful to find intermittent problems that would be difficult to find with a DMM. The DSO shows a wave pattern of the electrical voltage or current present in the circuit being tested. This feedback to the technician can be valuable to determine the exact cause of the problem.

15. Verify correct operation of certified equipment.

Air conditioning (A/C) recovery/recycling equipment must have a United Laboratories (UL) approval. The current SAE standard for R12 equipment is J1990 and the current SAE standard for R134a equipment is J2788.

The refrigerant container specified by the recovery/recycling equipment manufacturer must be used in this type of equipment to ensure that the container has proper capacity.

The Federal Clean Air Act (CAA), Section 609, established the law that anyone who performs a service involving refrigerant in an automobile air conditioning system must be properly licensed.

16. Recycle or properly dispose of refrigerant.

If the moisture warning light is on during the recycling process, the refrigerant contains excessive moisture and the filter/cartridge in the recovery/recycling equipment must be changed. After recovery, many refrigerant recovery systems display the amount of refrigerant and oil recovered from the system. A multipass system may not complete all recycling stages before storing the refrigerant. UL approval is required of all recovery/recycle equipment. SAE standard J1990 and J2788 are also required. There are two types of UL-approved recycle equipment: single pass and multipass. A single-pass system will not remove all contaminants. A multipass system will, however, clean and dry refrigerant to standards for reuse before dispensing it.

17. Label and store refrigerant.

Refrigerant storage containers must be filled to 60 percent of their gross weight rating. Refrigerant storage containers must be evacuated to 17 in. Hg (43.9 kPa absolute) before refrigerant is placed in the container.

When recovered refrigerant is stored, a DOT4BW cylinder needs to be used. Refrigerant should never be stored in anything else. Stored refrigerant should be placed in a cool and dry area, away from any heat sources.

18. Test recycled refrigerant for non-condensable gases; identify refrigerant.

When a refrigerant container is checked for non-condensable gases, the container should be stored out of the presence of sunlight at 65°F (18°C) for 12 hours and a thermometer should be placed 4 inches (102 mm) from the container surface. If the container pressure is less than specified, the refrigerant is ready for use.

Before recycled refrigerant is tested for non-condensable gases, the portable refrigerant cylinder should be placed in a relatively cool area out of direct sunlight for no less than 12 hours. The ambient temperature may be as low as 65°F (18°C). Compare the temperature and pressure reading with an appropriate pressure limit chart.

A refrigerant identifier is used to determine the makeup of the refrigerant in a container or an A/C system. Refrigerants that are found to be not pure must be handled with separate equipment that is only used for this type of refrigerant. A technician should never recover mixed refrigerant into the normal A/C recovery equipment because these machines are not capable of separating the mixed refrigerants.

19. Identify the procedures and equipment necessary to service, diagnose, and repair A/C systems in hybrid/electric vehicles.

Vehicles that are propelled with hybrid or pure electric technology often have HVAC systems that are different than regular internal combustion vehicles. The following list contains some of the major points to remember when working on these vehicles.

- Electric compressors: The compressors are driven by 3 phase AC current that is supplied by the high voltage system on the vehicle. The high voltage system must be turned off prior to performing intensive repair work on the compressor. Orange wires are often used on the high voltage components. Always refer to the service procedure in the repair literature before working on a system like this. Electric compressors use an ester based refrigerant oil for lubrication.

- PTC heater technology: Special heaters can be used to create heat in the cabin without having hot coolant pumped into the cabin. Heat is supplied by sending current to these electric grids which causes them to get hot.
- Thermal storage containers: Coolant storage tanks can be used to store hot engine coolant after the engine is shut down. These containers can store coolant at an elevated temperature for several hours. This system is used to get hybrid engines up to normal operating temperature faster. Service technicians must be aware of this system when performing service on these vehicles.

20. Diagnose the cause of temperature control problems in the heater/ventilation system; determine needed repairs.

In most cases, problems with the heating system are problems with the engine's cooling system. Therefore most service work and diagnosis are done to the cooling system. Problems that pertain specifically to the heater are few: the heater control valve and the heater. Most often, if these two items are faulty, the engine's cooling system will be negatively affected. Both of these items are replaced, rather than repaired. In some cases, it is possible to make repairs to vacuum hose and electrical connections without removing the heater assembly. If it is necessary to remove the heater assembly, the cooling system must be drained before the heater core is removed.

When diagnosing a heating system that is inadequate, it is imperative that you follow a logical diagnostic routine. The first objective is to be sure there is adequate coolant in the cooling system. Next, verify that the engine reaches normal operating temperature. In many cases, you will find a thermostat that is staying open, and your diagnosis is complete. If you have a way to see coolant flow from the upper hose into the radiator, if the thermostat is open, you will see flow through the hose even when the engine is cool.

If the engine is reaching operating temperature and the heater output is not acceptable, the next step is to check to see that there is flow through the heater core. Some systems use a heater control valve. These valves may be controlled by vacuum or electricity. Be sure that the valve changes position, allowing coolant into the heater core. If that area is good or not applicable, check the inlet and outlet temperatures or the heater hoses at the core. If one is cold, the core is restricted. If both are hot and equal, the fins on the outside of the core may be restricted (usually from debris like leaves) or the blend door may not be opening to allow the blower to move air across the core. There will be some drop in outlet temperature if everything is working normally. Usually this would be 10 to 30°F. A drop more than that means the core is restricted. Likewise, if a system is always hot, even when on the cold setting, check the control valve, if there is one, and the blend door operation. ATC systems often use computer-controlled blend doors with position sensors that can fail.

21. Diagnose window fogging problems; determine needed repairs.

A leaking heater core is often the cause of a window fogging problem. Window fogging is the result of hot and humid air that condenses on the cooler glass. Cleaning a clogged evaporator/heater case drain tube often eliminates a window fogging problem. If the drain is clogged, a leaking heater core will allow engine coolant to accumulate in the housing. A clogged drain also allows condensed water vapor to accumulate. Both of these problems add moisture to the air, which tends to fog the window.

Window fogging may also be caused by a plugged air conditioning/heater case drain, which allows water to collect in the case. This water can become stagnant and produce a very pungent odor in the passenger compartment. The window will not fog due to a low coolant level.

A sticky film on the inside of the window may be caused by a coolant leak in the heater core. Under this condition, the coolant level in the radiator should be checked. The heater core should be repaired or replaced.

22. Perform engine cooling system tests (pressure, electrolysis, concentration, and contamination); determine needed repairs.

The cooling system is a vital part of any internal combustion engine. The cooling system is also the system that most internal combustion engines use to heat the cab. There are several tests that a technician needs to be able to perform: testing for leaks, testing the freeze protection, the pH level, and the DC voltage present in the coolant.

Leak testing the cooling system first involves performing a visual inspection as well as using your sense of smell. Check the coolant level in the radiator and the recovery tank while the engine is cool. If the level is low, then a leak is likely and needs to be found. Use a good flashlight or droplight to inspect all of the engine, cooling system and HVAC components for any traces of liquid coolant or coolant odor. A hand-operated pressure pump can be used to add pressure to the cooling system to assist in this leak inspection process. Be sure to not pressurize the system above the rating of the radiator cap to prevent creating a leak that was not there before the inspection. If the leak is very challenging to find, then coolant dye can be added to the radiator. A black light is helpful to exaggerate the color of the dye during the inspection process.

The freeze protection of the coolant can be checked by using a hydrometer, a refractometer, or test strips. The hydrometer and refractometer both weigh the antifreeze concentration level and give feedback on what the freeze point is. Using a test strip to check freeze protection involves dipping the strip into the coolant and then comparing the color to a chart. It is a good idea to perform one of these checks during the fall and winter months in order to advise customers of potential coolant services.

A pH meter or test strips can be used to check the pH level of the engine coolant. The general specification for quality coolant is a pH between 7.5 and 10. A lower level indicates an acidic and corrosive characteristic that will damage fragile components. A high pH causes aluminum to corrode and build up in the engine.

The chemical makeup and operating characteristics of the cooling system cause it to sometimes build up a voltage. There are two reasons that voltage could be present in the coolant. One possibility is a poor electrical accessory ground on the vehicle. Occasionally, an accessory with a poor ground will use the cooling system as a path for electrical flow. This type of "dynamic" coolant voltage would be present with the engine and accessories turned on. A quality DVOM is needed to test the DC voltage of the coolant. If the voltage is 0.3 volts or higher, then look for a poor ground. Another circumstance that could create DC voltage in the coolant is increased acidity in the coolant. When this happens, the cooling system has some characteristics of a DC battery. This "static" coolant voltage would be present when the engine and accessories are turned off. The specification for this test is 0.2 volts DC. If the "static" voltage is too high, then it is advisable to drain and fill the coolant.

23. Inspect and replace engine coolant hoses and pipes.

Check cooling system hoses for loose clamps, leaks, and damage. Hoses with cracks, abrasions, bulges, or swelling, or that crunch when you squeeze them must be replaced.

Care must be taken when removing hoses to avoid damage to radiator and heater core fittings. If the hose is to be replaced, it is wise to make a longitudinal cut in the hose and peel it off the fitting. If the hose is good and will be reused, it must be gently loosened with a tool designed for it.

24. Inspect, test, and replace radiator, pressure cap, coolant recovery system, and water pump.

There are three basic types of radiator caps: the constant pressure type, pressure vent type, and closed system type. The closed system type radiator cap is the one found on today's cars. The others are found on older model vehicles. The constant pressure type has a lower seal or pressure valve that is held closed until the coolant gets hot enough to build enough pressure to open the valve within the preset pressure range. The pressure vent type cap is similar to the constant pressure type, however, it has a vacuum-release valve that is opened by a weight and is kept open to the atmosphere until the pressure is great enough to move the weight and close the valve. This prevents atmospheric pressure from entering the radiator. Like the constant pressure cap, this cap opens to release pressure when it builds to the specified amount.

The closed system type works in the same way as the constant pressure cap, except that it is designed to keep the radiator full at all times. When the specified pressure is reached, some coolant is released into the recovery tank. When there is a vacuum in the radiator (caused by less coolant), the vacuum is used to pull coolant from the recovery tank. These radiator caps are not designed to be removed for coolant checks. Coolant is checked and fluid is added through the recovery tank.

Water pumps can be driven by gears, chains, timing belts, accessory belts, or electricity. A careful inspection of the water pump should be part of a normal maintenance routine. All mechanical water pumps have bearings that will produce an audible knocking noise when they are getting close to complete failure. A stethoscope or other listening device can be used to isolate this problem. Another diagnostic routine for water pumps is to check the inspection hole for leaks. Any coolant or coolant residue in this location should be noted and used as evidence for recommending a new water pump.

25. Inspect, test, and replace thermostat, thermostat by-pass, and housing.

The thermostat regulates minimum engine temperature by limiting the amount of coolant flow to the radiator during engine warm-up and in cold weather operation. Computer-controlled vehicles will remain in a cold fuel strategy if the engine does not reach full operating temperature.

When the thermostat is closed, by-pass hoses route water throughout the engine to ensure even temperature throughout it until it reaches a temperature that the radiator then takes over those duties. By-pass hoses are often overlooked during hose replacement procedures and are often the source of hard-to-locate leaks.

Most thermostats have a housing that attaches them to the engine and facilitates the attachment of a radiator hose. These housings can be made of cast iron, aluminum, stamped steel, and plastic. They are often the site of small leaks that eventually corrode the housing, creating bigger leaks. When inspecting them, look for pitting or warpage.

It used to be a common practice to file the housing. This was the cause of many thermostat failures because the thermostat usually fits in a recess and the housing provides clamping force. If the recess is in the housing, the thermostat may be clamped too tight and distort, causing it to drag or stick, resulting in overheating or slow warm-up. To avoid the potential need of buying an engine, be sure to check this area carefully.

26. Identify, inspect, recover coolant; flush and refill system with proper coolant; bleed system as necessary.

Cooling system service is a topic approached differently by technicians in different repair venues. Included here is information that should be generic to all repair technicians. Manufacturer-specific items should not appear in the test unless they are considered industry standard. It is very important to keep this in mind when taking the test.

When testing coolant, there are many methods available to reach the same end. Information needed includes the protection levels of the coolant for freezing, boiling, pH, corrosion, and in some vehicles, nitrites. For the most part, freezing and boiling protection are linked. Most manufacturers agree that a mixture of 50 percent water to 50 percent coolant provides the best component protection. Most also agree that care should be taken to be sure the correct coolant in the vehicle is used without mixing coolant types or changing to one not designed for the vehicle.

pH is a measurement of the acidity or alkaline qualities of the coolant. As coolant becomes older it drops toward the acid end of the pH scale. Most vehicles manufactured in Asia aim for around 7 to 9 and most vehicles manufactured in America and Europe aim for 8 to 9.5 on the scale. Low pH readings can be caused by a deteriorated antifreeze condition or a water-heavy blend, since water is more acidic than a coolant mix. Very high numbers can be caused by over-adding antifreeze or corrosion packages during service.

Corrosion protection is added when servicing the cooling system with recovery/ recycling equipment, and it is in the antifreeze to begin with. This is a difficult area to test and a point of contention with manufacturers who do not support coolant recycling.

The last area that will become more critical as more diesel vehicles enter the consumer market is nitrites. When out of balance they cause small bubbles to collect on castings while the engine is running. These bubbles act like little cutters over time and carve into the casting. The vibration inherent in the diesel combustion process has been known to cause bubbles in the system that create leaks in cylinder walls. There are test strips available that detect the level of nitrites and currently only a couple of manufacturers have specifications for them for their light duty diesel trucks.

There are many compounds and chemical additives being used with coolant in today's marketplace. It is advisable to always use the correct type of antifreeze when servicing late-model vehicles. It is recommended to consult the manufacturer recommendations before purchasing replacement coolant.

The coolant must be captured as it is drained out of the vehicle to prevent the coolant from entering a drain or the ground. Used coolant should either be recycled on-site or removed by a hazardous materials hauler.

Refilling the cooling system is sometimes challenging on late-model vehicles. Some manufacturers have special funnels that can be used to help prevent air from entering the system. Many manufacturers install bleeder valves on the engine to assist in bleeding the air from the cooling system during refill. See task A.22 for more information on how to test the cooling system.

27. Inspect, test, and replace fan (both electrical and mechanical), fan clutch, fan belts, fan shroud, and air dams.

Inspect serpentine belts for missing ribs, wear on outside edges, cracks closer than 2 inches apart, and excessive glazing. Any of these require immediate replacement. V-belts with cracks or excessive glazing on the surfaces require replacement. Many vehicles are equipped with plastic tensioner and idler pulleys. It is very important to inspect these for wear. They will allow belt tension to drop or even damage the belt when they wear. Belt tensioners have positioning marks, which will indicate if the belt has reached a point where it is not under proper tension.

Before testing electric cooling fans, it is very important to check the wiring diagram and keep it handy. Many cooling fans simply use a thermal switch to turn them on and a relay to bypass that switch when the air conditioner is turned on. More sophisticated systems use the coolant temperature sensor and the powertrain control module to operate the fan at varying speeds for different loads. They will receive an A/C command from the control head or one of the switches under the hood. The PCM may even vary speed based on system pressure.

The mechanical fan systems used on air conditioned vehicles almost always incorporate a fan clutch. Fan clutches come in two varieties. The first and most common is the thermostatic clutch that locks up the fan more as the engine temperature rises. The other type is the centrifugal fan clutch that tries to maintain a consistent speed regardless of engine speed. The two most common fan clutch failures are no fan clutch engagement at all or a completely locked up clutch. The first will result in poor air conditioner efficiency and overheating in traffic. The second will result in a complaint that the vehicle sounds like an airplane and that there is a loss of power, as it can take up to an additional 17 horsepower to drive a multi-blade fan in a fixed mode at highway speeds. Inspect fan clutches for hydraulic leaks, looseness on the shaft, and damaged blades on the fans. Many of the plastic fan blades in use will develop cracks that can cause the blade to break and hit the radiator core.

The purpose of a fan shroud is to allow a round fan to create a low-pressure zone behind the entire radiator core. The purpose of an air dam is to create a high-pressure zone in front of the radiator. Both components encourage air flow through the radiator under different circumstances. The fan shroud supports good air flow at low speeds and the air dam supports good air flow at highway speeds. Keep in mind that a vehicle that overheats at highway speeds may have a problem with the air dam and a vehicle that overheats in traffic could have a problem with the fan shroud. The most common problems are that they are broken or missing

28. Inspect, test, and replace heater coolant control valve (manual, vacuum, and electrical types) and auxiliary coolant pump.

The heater control valve (sometimes called the water flow valve) controls the flow of coolant into the heater core from the engine. In a closed position, the valve allows no flow of hot coolant to the heater core, keeping it cool. In an open position, the valve allows heated coolant to circulate through the heater core, maximizing heater efficiency.

Cable-operated valves are controlled directly from the heater control lever on the dashboard. Thermostatically controlled valves have a liquid-filled capillary tube located in the discharge airstream off of the heater core. This tube senses air temperature and the valve modulates the flow of water to maintain a constant temperature, regardless of engine speed or temperature.

Most heater valves utilized on today's cars are vacuum operated. These valves are normally located in the heater hose line or mounted directly in the engine block. When a vacuum signal reaches the valve, a diaphragm inside the valve is raised, either opening or closing the valve against an opposing spring. When the temperature selection on the dashboard is changed, vacuum to the valve is vented and the valve returns to its original position. Vacuum-actuated heater control valves are normally either open or closed designs. Some vehicles don't use a heater control valve; rather, a heater door controls how much heat is released into the passenger compartment from the heater core.

With cable-operated control valves, check the cable for sticking, slipping (loose mounting bracket), or mis-adjustment. With valves that are vacuum operated, there should be no vacuum to the valve when the heater is on (except for those that are normally closed and need vacuum to open).

On late-model vehicles, heater control valves are typically made of plastic for corrosion resistance and light weight. These valves feature few internal working parts and no external working parts. With the reduced weight of these valves, external mounting brackets are not required.

If the manual coolant control valve is incorrectly adjusted and does not close completely, hot coolant will be allowed to pass through the heater core when the A/C is on. If the coolant control valve does not open completely, it will not allow enough hot coolant to pass through the heater core to properly heat the air in heat mode.

Rust discoloration at the heater control valve may indicate a malfunctioning valve. A defective heater control valve may result in a poor cooling condition by allowing hot coolant to enter the heater core. The heater control valve is used to control the flow of heated coolant through the heater core. The hot coolant is often used to "temper" the air for temperature control and to help maintain a desired in-car relative humidity. Heater coolant control valves may be electronically, mechanically, or vacuum controlled.

Auxiliary water pumps are used on some vehicles with rear climate control, as well as on some hybrid vehicles. These pumps typically operate on 12 volts DC received from a relay. These pumps can be inspected for leaks visually or by adding pressure with a pressure tester. A scan tool is often helpful when testing these pumps for proper running operation. A scan tool will display trouble codes as well as run functional tests on the pump.

29. Inspect, flush, and replace heater core.

The heater core is the main heat exchanger for the heating system. The two problems that can occur in heater cores are either leaks or restrictions. A leaking heater core will typically cause a distinct coolant odor in the cab, as well as windshield fogging when the defroster is turned on. If the heater core leak is large, then coolant will likely run out of the HVAC drain tube onto the ground. The HVAC box will have to be opened up to replace the heater core. On many vehicles, the HVAC box has to be removed from the vehicle before it can be opened to access the heater core. Care should be taken when performing this tedious job to prevent dash and instrument panel damage.

A gurgling noise in the heater core may be caused by a low coolant level in the cooling system or a restricted heater core. A low coolant level will allow too much air in the cooling system. The excessive air in the system will mix with the coolant and create the gurgling noise. A restricted heater core will also cause a gurgling noise from the coolant passing through the restricted area.

A restricted heater core will cause a complaint of the heater not getting hot enough. The thermostat should first be checked for correct operation to make sure that the engine is reaching normal operating temperature. If the thermostat is operating normally, then check the temperature drop across the inlet and outlet heater hoses. This test is performed on a fully warm engine with the blower on high and the control head set to full heat. There should be less than 15 degrees F drop across the heater core. If the temperature drop is greater than 15 degrees F, then the core will either need to be flushed or replaced. It is advisable to warn the customer that flushing the heater core could cause a leak to occur. If the customer approves of flushing the core, then proceed with a reverse flush procedure.

Air pockets have become more of a problem on late-model vehicles whenever the cooling system is serviced. Many engines have air bleeders installed at various locations to assist in purging the air from the cooling system. Always follow manufacturers' recommendations when refilling cooling systems.

B. Refrigeration System Component Diagnosis and Repair (10 Questions)

1. Compressor and Clutch

1. Diagnose A/C system problems that cause the protection devices (pressure, thermal, and electronic controls) to interrupt system operation; determine needed repairs.

An A/C pressure switch typically operates by opening and closing a set of electrical contacts. To diagnose these types of switches, a technician can check them with an ohmmeter to see if the resistance is high or low. The key is to make sure that the switch is opening and closing.

A system that is low on refrigerant will typically "short cycle," which means the compressor will turn on and off every few seconds. The low pressure switch is the device that causes the compressor to operate this way. See Task B.1.2 for more details.

2. Inspect, test, and replace A/C system pressure, thermal, and electronic protection devices.

A/C systems use various types of pressure and thermal switches and sensors to maintain correct operation under a variety of conditions. The table below lists the various devices and key characteristics

COMPONENT	ALTERNATIVE NAME	CHARACTERISTICS	# OF WIRES
Low-pressure switch	Pressure cycling switch	Opens under low A/C system pressure (opens at 15-25 psi; closes at 35-40 psi)	2
High-pressure switch	Pressure cutoff switch	Opens under high A/C system pressure (opens at 450-490 psi; closes at 250-275 psi)	2
Dual pressure switch	Binary pressure switch	Opens at high pressure and opens at low pressure. Located in high side.	3
Pressure sensor	A/C pressure transducer	Located in high side. Senses pressure in the high side by varying voltage level.	3
Located in high side. Senses pressure in the high side by varying voltage level.	Thermostatic switch	Opens when cold temperatures are experienced in evaporator.	2
Evaporator temperature sensor	Thermistor, fin temperature sensor	Senses temperature in evaporator. Variable signal.	2 or 3
Thermal limiter		Interrupts power to compressor when temperature of the compressor gets too high. Located on the compressor.	

Many manufacturers use the electrical devices in the table. Testing these devices is typically fairly simple. All pressure and thermal switches operate by opening and closing electrical contacts and the variables change. One way to test these items is to use an ohmmeter to watch the contacts open and close. When the contact is open, the ohmmeter should read out of limits (OL); when the switch is closed, the ohmmeter should read very close to zero ohms. Another way to test switches is to by-pass them with a fused jumper wire. To do this, simply disconnect the switch and then connect a fused jumper wire to the connector. If the circuit begins operating normally, then the switch that is being by-passed is open. One word of caution: Before by-passing a low-pressure switch, make sure that the refrigerant system is not totally empty. This can be checked by connecting a manifold set to test the static pressure.

The testing procedure is different for pressure or temperature sensors. These devices operate by varying a voltage signal as pressure or temperature changes. A digital voltmeter can be used to monitor signal voltage on these devices. In addition, a scan tool is also a very useful diagnostic tool when testing sensors. The scan tool will display the live data from each sensor as well as any trouble codes that might be present.

3. Inspect, adjust, and replace A/C compressor drive belts, pulleys and tensioners.

Always use the exact replacement size of belt. The size of a new belt is typically given, along with the part number, on the belt container. After replacing a belt, make sure it is adjusted properly. Some engines have an adjusting bolt that can be tightened to bring the belt tension to specifications. On other engines, it may be necessary to use a pry bar to move an accessory enough to meet tension specs. Be careful not to damage the part you are prying against. The belt's tension should be checked with a belt tension gauge.

When installing a serpentine belt, make sure it is fed in and around the accessories properly. Service manuals and under-hood decals show the proper belt routing. Also make sure the belt tensioner is working properly. After any drive belt has been installed with the correct tension, tighten any bolts or nuts that were loosened to move the belt. A loose A/C compressor belt will likely produce a squealing noise when the compressor clutch is engaged.

Pulleys can become worn and cause belt misalignment. Frequently, pulleys that run against the back of serpentine belts are made of plastic and will show wear. V-belt pulleys will also wear and result in belt noise on correctly tensioned new belts. A visual inspection will usually identify worn, bent, or misaligned pulleys. Replacement should follow manufacturers' recommended procedures and specifications.

4. Inspect, test, service, and replace A/C compressor clutch components or assembly.

Compressors are equipped with an electromagnetic clutch as part of the compressor pulley assembly. It is designed to engage the pulley to the compressor shaft when the clutch coil is energized. The purposes of the clutch are to transmit power from the engine to the compressor and to provide a means of engaging and disengaging the refrigeration system from engine operation.

The clutch is driven by power from the engine's crankshaft, which is transmitted through one or more belts to the pulley, which is in operation whenever the engine is running. When the clutch is engaged, power is transmitted from the pulley to the compressor shaft by the clutch drive plate. When the clutch is not engaged, the compressor shaft does not rotate, and the pulley freewheels.

The clutch is engaged by a magnetic field and disengaged by springs when the magnetic field is broken. When the controls call for compressor operation, the electrical circuit to the clutch is completed, the magnetic clutch is energized, and the clutch engages the compressor. When the electrical circuit is opened, the clutch disengages the compressor.

Two types of electromagnetic clutches have been in use for many years. Early model air conditioning systems used a rotating coil clutch. The magnetic coil, which engages or disengages the compressor, is mounted within the pulley and rotates with it. Electrical connections for the clutch operation are made through a stationary brush assembly and rotating slip rings, which are part of the field coil assembly. This older rotating coil clutch, now in limited use, has been largely replaced by the stationary coil clutch. The compressor clutch coil can be checked with an ohmmeter. Service manuals will give the acceptable resistance for the field coil of the clutch assembly.

With the stationary coil, wear has been measurably reduced, efficiency increased, and serviceability made much easier. When the driver first energizes the air conditioning system from the passenger compartment dashboard, the pulley assembly is magnetized by the stationary coil on the compressor body, thus engaging the clutch to the clutch hub that is attached to the compressor shaft.

This activates the air conditioning system. Depending on the system, the magnetic clutch is usually pressure controlled to cycle the operation of the compressor (depending on system temperature or pressure). In some system designs, the clutch might operate continually when the system is turned on. With stationary coil design, service is not usually necessary except for an occasional check on the electrical connections.

Nearly all clutch assemblies have a clearance spec for the distance between the clutch and the pressure plate. This clearance is measured with a feeler gauge. If the clearance is too great, the clutch may slip and cause a scraping or squealing noise. If the clearance is insufficient, the compressor may run when not electrically activated and the clutch may chatter at all times.

2012 © Delmar, Cengage Learning

5. Identify required lubricant type; inspect and correct level in A/C compressor.

Normally the only source of lubrication for a compressor is the oil mixed with the refrigerant. Because of the loads and speeds at which the compressor operates, proper lubrication is a must for long compressor life. The refrigerant oil required by the system depends on a number of things, but it is primarily dictated by the refrigerant used in the system. R-12 systems use a mineral oil. Mineral oil mixes well with R-12 without breaking down. Mineral oil, however, cannot be used with R-134a. R-134a systems require a synthetic oil, polyalkeline glycol (PAG). There are a number of different blends of PAG oil; always use the one recommended by the vehicle manufacturer or compressor manufacturer. Failure to use the correct oil will cause damage to the compressor.

Generally, compressor oil level is checked only where there is evidence of a major loss of system oil that could be caused by a broken refrigerant hose, severe hose fitting leak, badly leaking compressor seal, or collision damage to the system's components.

When replacing refrigerant oil, it is important to use the specific type and quantity of oil recommended by the compressor manufacturer. If there is a surplus of oil in the system, too much oil circulates with the refrigerant, causing the cooling capacity of the system to be reduced. Too little oil results in poor lubrication of the compressor. When there has been excessive leakage or it is necessary to replace a component of the refrigeration system, certain procedures must be followed to assure that the total oil charge in the system is correct after leak repair or the new part has been installed. When the compressor is operated, oil gradually leaves the compressor and is circulated through the system with the refrigerant. Eventually a balanced condition is reached in which a certain amount of oil is retained in the compressor and a certain amount is continually circulated. If a component of the system is replaced after the system has been operated, some oil goes with it. To maintain the original total oil charge, it is necessary to compensate for this by adding oil to the new replacement part. Because of the differences in compressor designs, be sure to follow the manufacturer's instructions when adding refrigerant oil to their unit.

When you replace an A/C compressor, the recommended procedure is to measure the oil recovered from the system while discharging. Drain and measure the oil left in the old compressor. Drain the new compressor and refill it with the same amount of oil removed from the system and old compressor. If 2 oz. (59 ml) of oil are recovered from the system and 2 oz. (59 ml) of oil drained from the old compressor, drain the replacement compressor and then add 4 oz. (118 ml) of oil to the compressor before installation.

A small amount of lubricant is lost from the compressor due to circulation through the system. The lubricant must be drained from some compressors and measured in a beaker. The lubricant used in automotive air conditioning systems is a non-foaming, sulfur-free grade specially formulated for use in certain types of air conditioning systems.

6. Inspect, test, service or replace A/C compressor, mounting, and fasteners.

The compressor is the heart of the automotive air conditioning system. It separates the high-pressure and low-pressure sides of the system. The primary purpose of the unit is to draw the low-pressure and low-temperature vapor from the evaporator and compress this vapor into high-temperature, high-pressure vapor. This action results in the refrigerant having a higher temperature than surrounding air and enables the condenser to condense the vapor back to a liquid. The secondary purpose of the compressor is to circulate or pump the refrigerant through the A/C system under the different pressures required for proper operation.

A piston compressor can have its pistons arranged in an in-line, axial, radial, or V design. It is designed to have an intake stroke and a compression stroke for each cylinder. On the intake stroke, the refrigerant from the low side of the system (evaporator) is drawn into the compressor. The intake of refrigerant occurs through intake reed valves. These one-way valves control the flow of refrigerant vapors into the cylinder. During the compression stroke, the vaporous refrigerant is compressed. This increases both the pressure and the temperature of the heat-carrying refrigerant. The outlet or discharge side reed valves then open to allow the refrigerant to move to the condenser. The outlet reed valves are the beginning of the high side of the system. Reed valves are made of spring steel, which can be weakened or broken if improper charging procedures are used, such as liquid charging with the engine running.

A common variation of a piston-type compressor is the variable displacement compressor. These compressors not only act as a regular compressor but they also control the amount of refrigerant that passes through the evaporator. The pistons are connected to a wobble-plate. The angle of the wobble-plate determines the stroke of the pistons and is controlled by the difference in pressure between the outlet and inlet of the compressor. When the stroke of the pistons is increased, more refrigerant is being pumped and there is increased cooling.

The rotary vane compressor does not have pistons. It consists of a rotor with several vanes and a carefully shaped housing. As the compressor shaft rotates, the vanes and housing form chambers. The refrigerant is drawn through the suction port into these chambers, which become smaller as the rotor turns. The discharge port is located at the point where the gas is completely compressed. No sealing rings are used in a vane compressor. The vanes are sealed against the housing by centrifugal force and lubricating oil. The oil sump is located on the discharge side, so the high-pressure tends to force it around the vanes into the low-pressure side. This action ensures continuous lubrication. Because this type of compressor depends on a good oil supply, it is subject to damage if the system charge is lost. A protection device is used to disengage the clutch if pressure drops too low.

The scroll-type compressor has a movable scroll and a fixed or non-movable scroll that provide an eccentric-like motion. As the compressor's crankshaft rotates, the movable scroll forces the refrigerant against the fixed scroll and toward the center of the compressor. This motion pressurizes the refrigerant. The action of a scroll-type compressor can be compared to that of a tornado. The pressure of air moving in a circular pattern increases as it moves toward the center of the circle. A delivery port is positioned at the center of the compressor and allows the high-pressure refrigerant to flow into the air conditioning system. This type of compressor operates more smoothly than other designs.

A compressor noise problem is often difficult to localize. When the compressor is operating, a loose compressor mount or bracket may vibrate, giving the impression that the compressor itself is somehow defective. This type of noise usually stops when the compressor is disengaged. It is very critical to torque the compressor fasteners to the correct value to keep them from getting too tight or leaving them too loose.

If the compressor mount is loose, this can cause the compressor to move and make noise. If the A/C system does not have any refrigerant in it, the compressor clutch will not engage and therefore will not make any noise.

2. Evaporator, Condenser, and Related Components

1. Inspect, repair or replace A/C system mufflers, hoses, lines, filters, fittings, and seals.

An air conditioning system is divided into two sides: the high side and the low side. High side refers to the side of the system that is under high-pressure and high temperature. Low side refers to the low-pressure, low-temperature side of the system.

All the major components of the system have inlet and outlet connections that accommodate either flare or O-ring fittings. The refrigerant lines that connect between these units are made up of an appropriate length of hose or tubing with flare or O-ring fittings at each end as required. In either case, the hose or tube end of the fitting is constructed with sealing beads to accommodate a hose or tube clamp connection.

There are three major refrigerant lines. Suction lines are located between the outlet side of the evaporator and the inlet side or suction side of the compressor. They carry the low-pressure, low-temperature refrigerant vapor to the compressor where it again is recycled through the system. Suction lines are always distinguished from the discharge lines by touch and size. The suction line is cold while the system is operating and is also larger in diameter than the liquid line because refrigerant in a vapor state takes up more room than refrigerant in a liquid state.

The other types of refrigerant lines are the discharge and liquid lines. Beginning at the discharge outlet on the compressor, the discharge or high-pressure line connects the compressor to the condenser. The liquid lines connect the condenser to the metering device. Through these lines, the refrigerant travels in its path from a gas state (compressor outlet) to a liquid state (condenser outlet) and then to the inlet side of the expansion valve, where it vaporizes on entry to the evaporator. Discharge and liquid lines are always very warm to the touch and easily distinguishable from the suction lines.

Aluminum tubing is commonly used to connect air conditioning components where flexibility is not required. Where the line is subjected to vibrations, special rubber hoses are used. Typically the compressor outlet and inlet lines are rubber hoses with aluminum ends and fittings.

R-134a systems are required to be fitted with quick-disconnect fittings through the system. These also have hoses specially made for R-134a. They have an additional layer of rubber that serves as a barrier to prevent the refrigerant from escaping through the pores of the hose. Some late-model R-12 systems also use these barrier hoses to prevent the loss of refrigerant through the walls of the hoses.

Some refrigerant line connections are sealed with O-rings and retained with spring lock couplings. A special tool is used to release the spring lock couplings.

It is not necessary to replace the O-ring seals unless they are leaking. The accumulation of dirt around an A/C line connection is an indication of a leak in the system. When there is a leak in the system, dirt collects in the area of the leak due to some of the oil leaking out.

2. Inspect A/C condenser for proper air flow.

The condenser consists of coiled refrigerant tubing mounted in a series of thin cooling fins to provide maximum heat transfer in a minimum amount of space. The condenser is normally mounted just in front of the vehicle's radiator. It receives the full flow of ram air from the movement of the vehicle or air flow from the radiator fan when the vehicle is standing still.

The purpose of the condenser is to condense or liquefy the high-pressure, high-temperature vapor coming from the compressor. To do so, it must give up its heat. The condenser receives very hot (normally 200 to 400°F or 93 to 204°C), high-pressure refrigerant vapor from the compressor through its discharge hose. The refrigerant vapor enters the inlet at the top of the condenser, and as the hot vapor passes down through the condenser coils, heat (following its natural tendencies) moves from the hot refrigerant into the cooler air as it flows across the condenser coils and fins. This process causes a large quantity of heat to be transferred to the outside air and the refrigerant to change from a high-pressure hot vapor to a high-pressure warm liquid. This high-pressure warm liquid flows from the outlet at the bottom of the condenser through a line to the receiver/drier or to the refrigerant metering device if an accumulator instead of a drier is used.

In an air conditioning system that is operating under an average heat load, the condenser has a combination of hot refrigerant vapor in the upper two-thirds of its coils. The lower third of the coils contains the warm liquid refrigerant, which has condensed. This high-pressure, liquid refrigerant flows from the condenser and on toward the evaporator. In effect, the condenser is a true heat exchanger.

Ram air flow across the condenser is produced by vehicle movement or the action of the fan. A defective fan clutch generally causes the fan to run at maximum potential at all times, which will usually increase ram air flow. Reduced air flow across the condenser generally affects the ability of the condenser to cool the refrigerant, resulting in higher temperatures. The higher refrigerant temperatures increase suction pressure in the system, affecting system cooling abilities.

If the condenser air passages are severely restricted, this may result in high refrigerant system pressures and refrigerant discharge from the high-pressure relief valve. Restrictions will not allow enough air to cool down the condenser, causing higher temperatures and in turn higher pressures.

3. Inspect, test, and clean or A/C system condenser; check mountings and air seals.

The condenser is a vital heat exchanger that is typically located in front of the radiator. The job of the condenser is to allow the gaseous refrigerant to release its heat onto the fins and then into the surrounding air, which causes the refrigerant to change back into a liquid.

The condenser should be checked regularly for damaged fins as well as debris or dirt that would impede airflow. It is a good practice to routinely wash the surface of the condenser in order to assure proper heat transfer. Any time that any internal particles should get into the condenser, it should be flushed or replaced. Many late-model vehicles have condensers that use either serpentine or parallel flow designs. Both of these designs have very small passages that can easily get restricted and are extremely challenging to clear out.

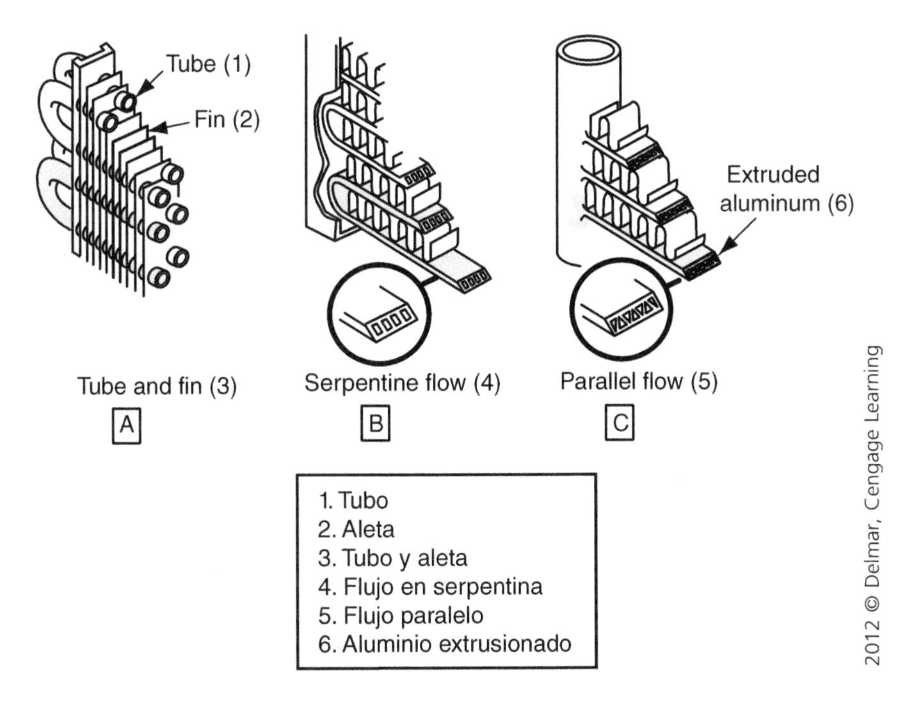

The correct amount of refrigerant oil should be installed into the condenser after being flushed or replaced. Most manufacturers recommend one to two ounces be added at this time. Always consult the proper service information when making this repair. All condensers should be securely mounted with the correct mounts and brackets. The air seals that route the airflow through the condenser are also vital and must be in place and not torn or damaged.

A temperature drop test is an effective method for testing condenser operation. This test is run with the A/C on maximum, the blower motor on high, and the doors open. An accurate contact thermometer is used to measure the condenser inlet temperature as well as the condenser outlet temperature. The range of temperature drop should be between 20°F and 50°F.

4. Inspect and replace receiver/drier; accumulator/drier, or dessicant

TXV-style A/C systems use a receiver/drier near the outlet of the condenser. Some manufacturers have combined the receiver dryer with the condenser as an assembly. This practice reduces the number of potential leak points on the system. Some of these combined units have removable plugs that allow the replacement of the dessicant bag. If there is no access point, then the condenser must be replaced in order to repair a problem with the receiver dryer.

As the refrigerant flows through an opening in the lower portion of the receiver, it is filtered through a mesh screen attached to a baffle at the bottom of the receiver. The desiccant in this assembly is to absorb any moisture present that might enter the system during assembly. These features of the assembly prevent obstruction to the valves or damage to the compressor.

Depending on the manufacturer, the receiver/drier may be known by other names such as filter or dehydrator. Regardless of its name, the function is the same. Included in many receiver/driers are additional features such as a high-pressure fitting, a pressure relief valve, and a sight glass for determining the state and condition of the refrigerant in the system.

The receiver/drier is often neglected when the A/C system is serviced or repaired. Failure to replace it can lead to poor system performance or replacement part failure. It is recommended that the receiver/drier and/or its desiccant be changed whenever a component is replaced, the system has lost the refrigerant charge, or the system has been open to the atmosphere for any length of time.

Orifice tube-style A/C systems use an accumulator/drier to perform the same types of jobs as the receiver/drier. The accumulator is connected into the low side, at the outlet of the evaporator. The accumulator also contains a desiccant and is designed to store excess refrigerant and to filter and dry the refrigerant. If liquid refrigerant flows out of the evaporator, it will be collected by and stored in the accumulator. The main purpose of an accumulator is to prevent liquid from entering the compressor.

5. Inspect, test, and replace expansion valve(s).

The refrigerant flow to the evaporator must be controlled to obtain maximum cooling, while ensuring complete evaporation of the liquid refrigerant within the evaporator. This is accomplished by a thermostatic expansion valve (TXV) or a fixed orifice tube.

The TXV is mounted at the inlet to the evaporator and separates the high-pressure side of the system from the low-pressure side. The TXV regulates refrigerant flow to the evaporator to prevent evaporator flooding or starving. In operation, the TXV regulates the refrigerant flow to the evaporator by balancing the inlet flow to the outlet temperature.

Both externally and internally equalized TXVs are used in air conditioning systems. The only difference between the two valves is that the external TXV uses an equalizer line connected to the evaporator outlet line as a means of sensing evaporator outlet pressure. The internal TXV senses evaporator inlet pressure through an internal equalizer passage. Both valves have a capillary tube to sense evaporator outlet temperature. The tube is filled with a gas, which, if allowed to escape due to careless handling, will ruin the TXV.

During stabilized conditions, the pressure on the bottom of the expansion valve diaphragm becomes equal to the pressure on the top of the diaphragm. This allows the valve spring to close the valve. When the system is started, the pressure on the bottom of the diaphragm drops rapidly, allowing the valve to open and meter liquid refrigerant to the lower evaporator tubes where it begins to vaporize.

Compressor suction draws the vaporized refrigerant out of the top of the evaporator at the top tube where it passes by the sealed sensing bulb. The bottom of the valve diaphragm internally senses the evaporator pressure through the internal equalization passage around the sealed sensing bulb. As evaporator pressure is increased, the diaphragm flexes upward, pulling the pushrod away from the ball seat of the expansion valve. The expansion valve spring forces the ball onto the tapered seat, and the liquid refrigerant flow is reduced.

As the pressure is reduced due to restricted refrigerant flow, the diaphragm flexes downward again, opening the expansion valve to provide the required controlled pressure and refrigerant flow condition. As the cool refrigerant passes by the body of the sensing bulb, the gas above the diaphragm contracts and allows the expansion valve spring to close the expansion valve. When heat from the passenger compartment is absorbed by the refrigerant, it causes the gas to expand. The pushrod again forces the expansion valve to open, allowing more refrigerant to flow so that more heat can be absorbed.

A defective capillary tube would most likely cause the thermal expansion valve (TXV) to remain closed, causing poor cooling at all times. An overcharged system will cause higher suction pressures and poor cooling at all times. A restricted condenser passage will cause low cooling at all times. Some symptoms indicate moisture freezing in the TXV. When the TXV freezes, refrigerant flow is reduced and cooling is also reduced. As the TXV thaws, refrigerant flow is restored, and cooling is also restored until the TXV freezes again.

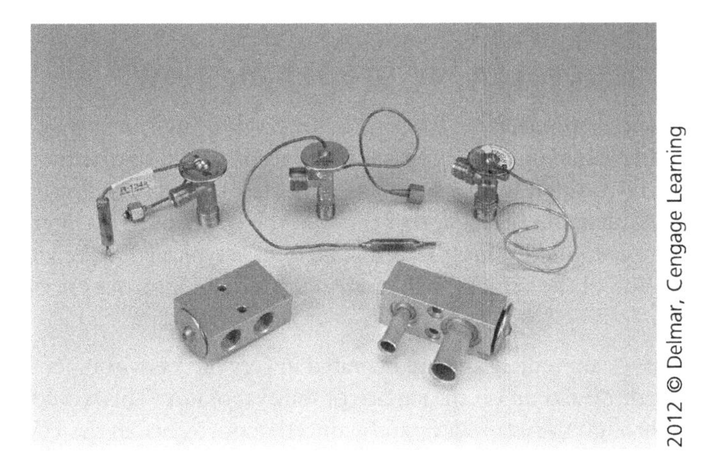

6. Inspect and replace orifice tube(s).

Like the thermostatic expansion valve, the orifice tube is the dividing point between the high- and low-pressure parts of the system. However, its metering or flow rate control does not depend on comparing evaporator pressure and temperature. It is a fixed orifice. The flow rate is determined by pressure difference across the orifice and by sub-cooling. Sub-cooling is additional cooling of the refrigerant in the bottom of the condenser after it has changed from vapor to liquid. The flow rate through the orifice is more sensitive to sub-cooling than to pressure difference.

There have been some advancements in orifice tubes in recent years. One advancement is the variable orifice valve and its ability to change its flow rate depending on the needs of the refrigerant system. These devices are used on vehicles that idle for long periods of time such as taxis and police cars. Another advancement that is growing in use is the electronic orifice tube. This device is a solenoid which can change its flow rate when the voltage signal changes. The PCM is the component that sends the signal to open and close it. These electronic devices can be diagnosed with scan tools and digital multimeters (DMM).

When using the special tool to remove an orifice tube, pour a small amount of refrigerant oil on top of the tube and engage the notch in the tool in the orifice tube. Hold the T-handle and rotate the outer sleeve to remove the orifice tube.

A restricted orifice tube may cause lower-than-specified low-side pressure. A restricted orifice tube may cause frosting of the tube because of the low temperatures caused by the low-pressure.

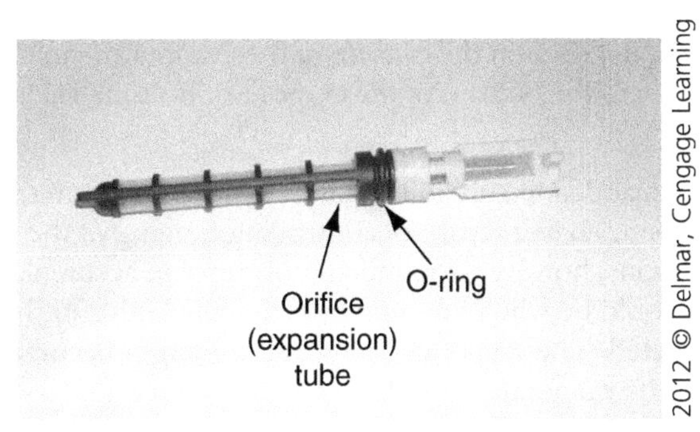

Variable orifice valves are used on some late-model vehicles. This type of orifice tube does change its size as the temperature load changes. This prevents the evaporator core from getting flooded during extended periods of vehicle idling. Police cars and taxi cabs are often used in this pattern.

7. Inspect A/C evaporator for proper air flow.

The air flow through the evaporator can be checked by monitoring the flow of air from the ducts. This should be done on each of the blower speeds to assure that the proper amount of air is being distributed from the system. Most late model vehicles are equipped with cabin air filters that clean the air as it flows through the duct box. This filter should be inspected if low air flow is found during this test. The filter is typically located in the duct box on the passenger side of the vehicle. See task A.10 for more information on cabin filters.

Vehicles that do not have cabin filters and are operated in very dirty environments can potentially accumulate dirt build up on the surface of the evaporator. This would cause a reduction of air flow through the evaporator and reduced performance in the HVAC system. The duct box would have to be opened in order to clean the evaporator surface manually.

Some manufacturers have had problems with the evaporator producing a musty smell when the HVAC system is operated. The cause of this bad smell is often from a moldy substance that appears on the surface of the evaporator core. This smell is often described as being like dirty gym socks. The repair procedure for this problem is to either replace the evaporator core or to spray a chemical on the surface that kills the material. Access to the evaporator core area is sometimes possible by removing the blower resistor or the blower motor. After the problem is corrected, it is advisable to operate the system on fresh air as much as possible to reduce the chance of this problem occurring again.

8. Inspect, test, clean, or replace evaporator(s).

The evaporator, like the condenser, consists of a refrigerant coil mounted in a series of thin cooling fins. It provides a maximum amount of heat transfer in a minimum amount of space. The evaporator is usually located beneath the dashboard or instrument panel.

Upon receiving the low-pressure, low-temperature liquid refrigerant from the thermostatic expansion valve or orifice tube in the form of an atomized (or droplet) spray, the evaporator serves as a boiler or vaporizer. This regulated flow of refrigerant boils immediately. Heat from the core surface is lost to the boiling and vaporizing refrigerant, which is cooler than the core, thereby cooling the core. The air passing over the evaporator loses its heat to the cooler surface of the core, thereby cooling the air inside the car. As the process of heat loss from air to the evaporator core surface is taking place, any moisture (humidity) in the air condenses on the outside of the evaporator core and is drained off as water. A drain tube in the bottom of the evaporator housing leads the water outside the vehicle.

This dehumidification of air is an added feature of the air conditioning system that adds to passenger comfort. It can also be used as a means of controlling fogging of the vehicle windows. Under certain conditions, however, too much moisture can accumulate on the evaporator coils. An example would be when humidity is extremely high and the maximum cooling mode is selected. The evaporator temperature might become so low that moisture would freeze on the evaporator coils before it can drain off.

Through the metering, or controlling, action of the thermostatic expansion valve or orifice tube, greater or lesser amounts of refrigerant are provided in the evaporator to adequately cool the car under all heat load conditions. If too much refrigerant is allowed to enter, the evaporator floods. This results in poor cooling due to the higher pressure (and temperature) of the refrigerant. The refrigerant can neither boil away rapidly nor vaporize. On the other hand, if too little refrigerant is metered, the evaporator starves. Poor cooling again results because the refrigerant boils away or vaporizes too quickly before passing through the evaporator.

The temperature of the refrigerant vapor at the evaporator outlet will be approximately 4 to 16°F higher than the temperature of the liquid refrigerant at the evaporator inlet. This temperature differential is the superheat that ensures that the vapor will not contain any droplets of liquid refrigerant that would be harmful to the compressor.

9. Inspect, clean, and repair evaporator housing and water drain.

A plugged evaporator case drain may cause window fogging, but this problem would not result in an oily film on the window. An oil film on the window may be caused by a refrigeration leak in the evaporator core that may allow some refrigerant oil to escape in the evaporator case.

The evaporator core should be removed from the case for proper cleaning. Some evaporators have no filter; therefore, airborne debris, such as lint, hair, and other contaminants, enters the evaporator case and clings to the wet evaporator core. It is nearly impossible to adequately clean an evaporator core without first removing it from the case. However, many late-model vehicles have a cabin air filter that will reduce the likelihood of debris getting into the evaporator core.

The evaporator water drain is an integral part of the evaporator housing assembly. A crack or break in the housing may reduce the cooling capacity of the air conditioner. If the evaporator (water) drain becomes clogged, it is easily cleaned with a stiff wire from the outlet in, provided you take care not to puncture the delicate evaporator core. If the case has a crack or break on the engine side of the firewall, excessive engine heat may enter the case and thereby reduce the efficiency of the A/C system.

10. Inspect, test, and replace evaporator pressure/temperature control systems and devices.

Evaporator controls maintain a backpressure in the evaporator. Because of the refrigerant temperature/pressure relationship, the effect is to regulate evaporator temperature. The temperature is controlled to a point that provides effective air cooling but prevents the freezing of moisture that condenses on the evaporator.

In this type of system the compressor operates continually when dash controls are in the air conditioning position. Evaporator outlet air temperature is automatically controlled by an evaporator pressure control valve. These types of valves throttle the flow of refrigerant out of the evaporator as required to establish a minimum evaporator pressure and thereby prevent freezing of condensation on the evaporator core.

In some refrigerant systems with an evaporator pressure regulator (EPR), pilot-operated absolute valve (POA), or suction throttling valve (STV) valve, the compressor runs continually in the A/C mode. Excessive pressure drop across the EPR valve indicates this valve is sticking closed.

For some time, the evaporator temperature regulator (ETR) and suction throttling valve (STV) were popular methods of evaporator temperature control, but these have not been in use for several years. All were used in the suction line from the evaporator to the compressor to maintain evaporator pressure, thereby maintaining its temperature.

11. Identify, inspect, and replace A/C system service valves and valve caps.

System service valves, which incorporate a service gauge port for manifold gauge set connections, are provided on the low and high sides of most A/C systems. When making gauge connections, purge the gauge lines first by cracking the charging valve and allowing a small amount of refrigerant to flow through the lines, and then connect the lines immediately.

R-12 systems have Schrader service valves in both the high and low portions of the system for test purposes. Closely resembling a tire valve, Schrader valves are usually located in the high-pressure line (from compressor to metering device) and in the low-pressure line (from evaporator to compressor) to permit checking of the high side and low side of the system. All test hoses have a Schrader core depressor in them. As the hose is threaded into the service port, the pin in the center of the valve is depressed, allowing refrigerant to flow to the manifold gauge set. When the hose is removed, the valve closes automatically.

R-134a service valves are the round quick-connect design. The service hoses for R-134a A/C equipment easily snap onto the service ports. The low-side service port is 13mm in diameter, and the high-side service port is 16mm in diameter. Once the service hoses are connected, the valves have to be turned clockwise in order to let refrigerant pass the one-way check valve on the connection.

After disconnecting gauge lines, check the valve areas to be sure the service valves are correctly seated and the Schrader valves are not leaking. This last step can save you from having to recharge a system for free. This is where an electronic leak detector can come in very handy. After servicing a system, just check the service ports for leaks before replacing the caps.

12. Inspect and replace A/C system high-pressure relief device.

The high-pressure relief valve is used to keep system pressures from reaching a point that may cause compressor lockup or other component damage because of excessive high pressures. When system pressures exceed a predetermined point, the pressure relief valve opens, allowing excessive system pressures to be reduced. A high-pressure relief valve discharges refrigerant at approximately 475 psi (3275 kPa).

If you are replacing the high-pressure relief valve, the system must be recovered before removing the valve. It is necessary to replace the high-pressure relief valve with one of the same relief pressure specifications unless the vehicle is going to be retrofitted to use R-134a refrigerant instead of R-12.

Some high-pressure relief devices are spring-loaded and self-resetting. Some high-pressure relief devices are made of fusible metal and are not self-resetting. Self-resetting high-pressure relief valves are usually found on the rear head of the compressor, while fusible metal high-pressure relief devices are often found on the receiver/drier.

C. Operating Systems and Related Controls Diagnosis and Repair (19 Questions)

1. *Electrical*

1. Diagnose the cause of failures in the electrical control system of heating, ventilating, and A/C (HVAC) systems; determine needed repairs.

Electrical problems can be classified as one of three types: shorts, opens, and high resistance. Each type affects the system differently and the problem source will be found at various locations. A short is no more than an unwanted parallel circuit. Sometimes the short is a connection to the power side of another circuit; at other times it is a connection to ground. In either case, a short will change the way the circuit operates. A short to ground before the load will cause a fuse or circuit breaker to blow. A short to ground after the load will not affect the circuit if it is switched on the power side. This type of short will cause the circuit to operate continually if the switch is on the ground side. An open circuit is caused by a break or disconnected wire or component. No current will flow through the circuit. Unwanted high-resistance problems reduce the current flow through the circuit. They also reduce the amount of voltage available to the circuit's components. Common high-resistance problems are caused by loose or disconnected grounds, loose terminals, corroded terminals, or frayed wire strands. Quality wire repairs using weather-sealed connections should always be performed on any suspected wiring problems.

2. Inspect, test, repair, and replace HVAC heater blower motors, blower motor speed controls, resistors, switches, relays/modules, wiring, and protection devices.

The blower motor/fan assembly is located in the evaporator housing. Its purpose is to increase airflow in the passenger compartment. The blower, which is basically the same type as those used in heater systems, draws warm air from the passenger compartment, forces it over the coils and fins of the evaporator, and blows the cooled, cleaned, and dehumidified air into the passenger compartment. The blower motor is controlled by a fan switch.

An open circuit at the blower switch ground or at the terminal causes the blower to be completely inoperative. An open resistor will usually cause the blower motor to operate only in high speed. When a relay is used in the blower circuit, it is typically energized for high speed due to the increased level of current flow on high. A faulty blower relay will typically cause the blower to work on the lower speeds, but not on high speed. A worn motor can sometimes cause a blown blower motor fuse due to the increased physical resistance. A current draw test can be used to determine if the blower motor is drawing excessive current.

To function properly, the blower motor requires a complete electrical circuit. It is generally connected to the negative side of the battery via the frame of the vehicle and to the positive side of the battery via insulated wires, speed resistors, the switch, the fuse or circuit breaker, and the ignition switch.

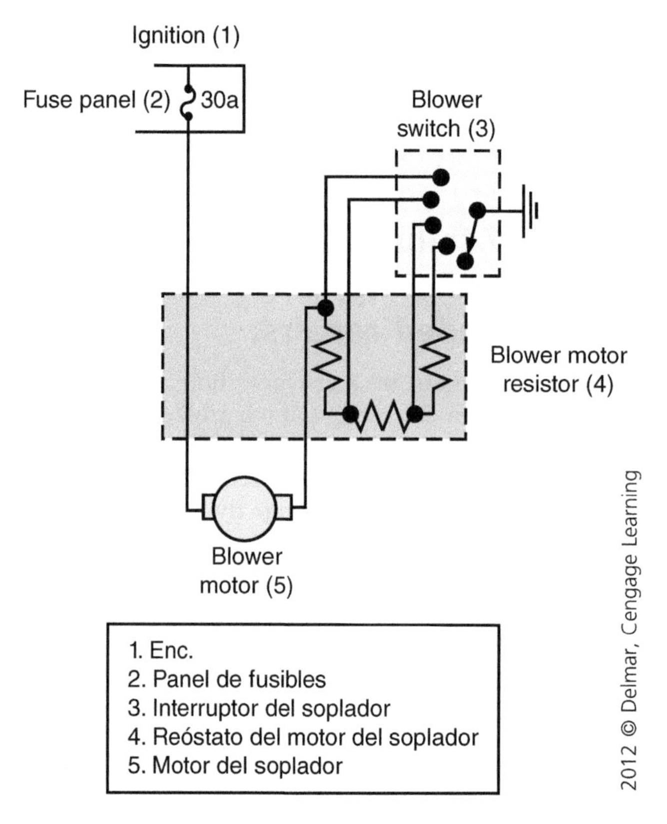

3. Inspect, test, repair, and replace A/C compressor clutch coil, relay/modules, wiring, sensors, switches, diodes, and protection devices.

Most late-model A/C systems have several controls intended to protect the compressor during extreme conditions, like stop-and-go traffic, and to provide extra power when needed. Nearly all R-134a systems incorporate some way to measure both excessive high-side pressure and low-side pressure for either compressor cycling or to shut the system down if a major loss of refrigerant occurs. See Task B.1.2 for further details about the sensors and switches used.

Relays are used in A/C systems to provide the PCM or other low amperage switches with a means to control functions like cooling fans, compressor clutch engagement, and/or wide-open throttle cutout used to drop out the compressor for additional power under load. Most late-model scan tools allow you to run a functional test on the A/C relay and clutch coil to see if both of these items will function.

Many A/C compressors have a diode wired in parallel with the clutch coil to act as a clamping device for the voltage spike that is created when the coil is de-energized. This clamping diode is connected in reverse bias to the normal power flow so it will not conduct electricity when the coil is energized. If this clamping diode shorts out, then the fuse that supplies the circuit will likely blow due to the high current flow. If this clamping diode fails out of limits (OL), then the voltage spike could cause the ECM to get hit with this high voltage. Testing a clamping diode requires a quality DVOM with a diode test function or an ohmmeter.

4. Inspect, test, repair, replace, and adjust A/C-related powertrain control systems and components.

It is sometimes necessary to alter the A/C operation under various engine conditions. The following is a list of components that can alter A/C operation:

- Throttle position sensor: A potentiometer that varies a voltage signal as the throttle or pedal angle changes. The A/C is typically turned off when high levels of throttle are seen.
- Engine temperature sensor: A thermistor that varies a voltage signal as the temperature of the engine changes. The A/C can be deactivated if the engine temperature gets too high.
- Power steering switch: A pressure switch that opens when high power steering pressure is experienced. The A/C can be deactivated when the power steering switch opens in parking maneuvers to prevent excessive load on the engine.

If any of the preceding components fail to operate correctly, trouble codes may set that can be retrieved with a scan tool. If no codes are present, then the scan tool can be used to monitor the voltage signals on each device to see if they change as the conditions change.

5. Inspect, test, repair, replace, and adjust load sensitive A/C compressor cut-off systems.

Some vehicles have a power steering cutoff switch to disengage the air conditioning compressor when the power steering requires maximum effort. Load-sensitive electrical switches include the low-pressure switch, high-pressure switch, pressure cycling, and power steering. Not all of these switches, however, are used on all vehicles. They are, for the most part, used to provide additional engine power when maximum power is required. The pressure switches are used to prevent compressor or component damage in the event of extremely high or low system pressures.

Some compressor clutch circuits contain a thermal limiter switch that senses compressor surface temperature. Some A/C compressor clutch circuits also contain a low-pressure and a high-pressure cutoff switch.

The following table describes the load-sensitive A/C-related components and methods for testing them. See Task B.1.2 for further details about the sensors and switches used.

COMPONENT	DESCRIPTION	TESTING METHOD	TESTING METHOD
Thermal limiter	Interrupts power to compressor when temperature of the compressor gets too high. Located on the compressor.	Use an ohmmeter to test the resistance with the circuit turned off.	Use a voltmeter to test the voltage in each side of the thermal limiter with the circuit energized.
Power steering switch	Opens when very high power steering pressure is present (especially, during parallel parking).	Use an ohmmeter to test the resistance with the circuit turned off.	Use a voltmeter to test the voltage in each side of the thermal limiter with the circuit energized.
Coolant sensor	Inhibits A/C operation under high engine temperatures.	Use a scan tool to read the live data from the coolant sensor or to retrieve trouble codes related to the sensor.	
Throttle position sensor (TPS)	Inhibits A/C operation under WOT (wide open throttle) conditions.	Use a scan tool to read the live data from the coolant sensor.	Use a scan tool to retrieve trouble codes related to the sensor.

6. Inspect, test, repair, and replace engine cooling/ condenser fan motors, relays/modules, switches, sensors, wiring, and protection devices.

Some cooling fan systems use a two-fan system: They have a low-speed fan and a high-speed fan. An electric cooling fan is usually wired through a relay because of the amount of current drawn to run an electric fan.

A scan tool is the tool of choice to interpret engine temperature data as well as A/C-related data that would affect the fan motor. The scan tool will display data and trouble codes as well as command output tests of various engine-driven items, such as the cooling fan.

Some late-model vehicles use a pulse width modulated (PWM) electric cooling fan motor. These fan motors are typically energized by some type of driver module that sends a varying voltage to the fan motor. These modules receive a PWM signal from the ECM/PCM and then they send a higher amperage PWM signal to the cooling fan motor. These fan motors have many different speeds available, depending on the heat load that is present.

A blown cooling fan fuse will cause the cooling fans to be inoperative at all times. If the system is equipped with high- and low-speed fan operations, an open circuit in one of the circuits may cause fan operation in only one speed.

7. Inspect, test, adjust, repair and replace climate control system electric actuator motors, relays/ modules, switches, sensors, wiring, and protection devices (including dual/multi-zone systems).

HVAC systems that use electric motors to move the various doors and ducts in the HVAC box typically use some type of feedback sensor to allow the controller to know the doors' position at all times. Most of these systems will use a potentiometer as the feedback device. The systems that operate this way also allow scan tools to view data streams and diagnostic codes, and run functional tests. The figure below shows communication flow among components.

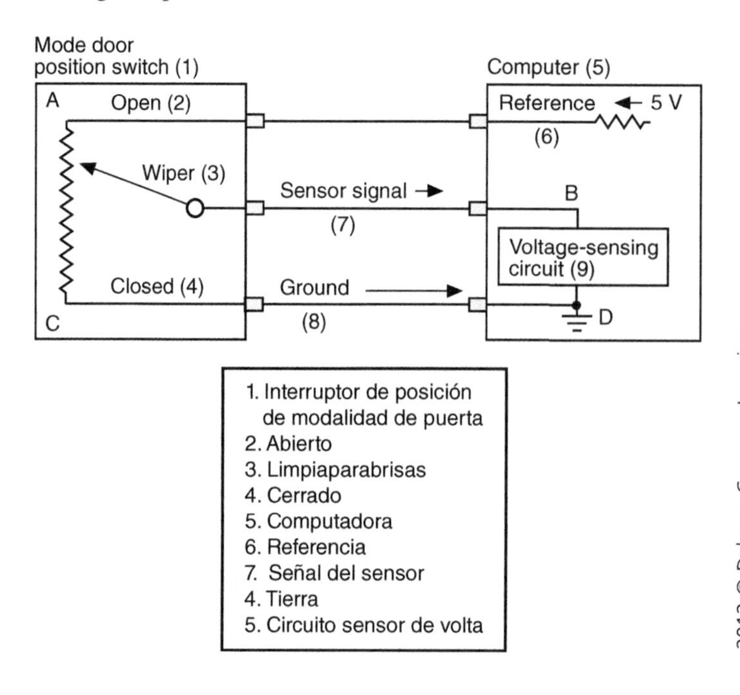

These actuators are usually serviced without having to remove the complete HVAC duct box. If a duct or door fails, then the complete HVAC duct box would have to be removed from the vehicle. After removing the duct box from the vehicle, the assembly can be opened up to gain access to the ducts and doors. Extreme care should be taken when reassembling these duct boxes to ensure the correct functionality when the repair is complete.

Replacing the electric motors on the HVAC system can vary in difficulty depending on the location of the motor. The HVAC system will usually need to be calibrated after replacing any of the electric motors. It is also necessary to calibrate some vehicles' HVAC doors each time the battery loses power or is disconnected.

Dual/multi-zone HVAC systems allow the driver, passenger, and even rear seat occupants to choose the temperature of the air that blows in their areas. These types of HVAC systems are always controlled by logic devices that look for inputs from the various locations. The air box for the driver and passenger is more complex due to the need for a blend door for both the driver and passenger. Calibration of these systems will need to take place after any component is replaced or disconnected for other service procedures. There are typically two ways to calibrate these systems. They can usually be calibrated by entering a manual self-diagnostic mode, which is done by pressing two of the control head buttons in a particular sequence. A scan tool is also capable of calibrating the HVAC doors.

8. Inspect, test, service, or replace HVAC panel assemblies.

To inspect and test the HVAC control panel, the technician needs to start the vehicle and get it up to operating temperature. Then all of the mode settings, temperature settings, and blower speeds should be tested to monitor for correct feedback. If any of the modes, temperatures, or blower settings do not work correctly, then further diagnosis will need to take place to see if the control head or other parts of the system are at fault.

Due to the fact that many HVAC control heads are electronic, it is advisable to disconnect the negative battery cable prior to removing the control head. After the panel is replaced, the technician will need to recalibrate the system either by entering self-diagnostics on the control head or by using a scan tool. All functions should be checked for proper operation after replacement of the control head.

2. *Vacuum/Mechanical*

1. Diagnose the cause of failures of the heating, ventilating, and A/C (HVAC) vacuum and mechanical control systems; determine needed repairs.

Problems in the HVAC vacuum system include leaking hoses, valves, controls, and actuators. The technician needs to use a good strategy when troubleshooting problems with these components. One useful tool is the technician's sense of hearing. Many times a vacuum leak can be heard as a hissing sound. Another quality tool that works well is a vacuum pump. The technician can isolate parts of the vacuum system and pinpoint the location of the loss of vacuum.

Problems that can occur in a mechanically operated HVAC system include cable problems, linkage problems, and binding door problems. The key to diagnosing these systems is to verify that the doors are moving to the desired positions. Again, the technician's sense of hearing works well as a diagnostic aid. The technician can usually hear the HVAC doors hitting the limits of travel. If the thud sound of hitting the travel limit is not heard, then an adjustment is likely needed.

2. Inspect, test, service, or replace HVAC control panel assemblies.

The HVAC control panel assembly is the input device that the driver or passenger uses to control the temperature, the location, and the speed of the air that is being discharged from the HVAC system. Some older vehicles had some serviceable components on the HVAC control panel such as the blower switch, mode switch, light bulbs, and vacuum switching valve. The control panel would have to be removed to service these items on the bench and then the panel would have to be reinstalled.

Many late-model vehicles have non-serviceable HVAC control panels that must be replaced as an assembly. It is wise to disconnect the battery before beginning the removal process on any late-model vehicle dash component. A scan tool can be used in the diagnosis of many HVAC-related problems on these late-model vehicles. Always use the available technology and service resources to your advantage during the troubleshooting process.

3. Inspect, test, adjust, and replace HVAC control cables and linkages.

Manual HVAC controls are sometimes used on HVAC systems. These systems use cables and linkages to move the mode, fresh air, and temperature doors. These systems can be identified by listening closely when moving the HVAC control head. A distinct feel and sound of moving doors can be sensed when manual controls are used. One key diagnostic step is to listen while moving the HVAC control head to the limits of the lever travel. A distinct sound should be heard as the doors hit each limit of travel. If this sound is not heard, then a cable or linkage adjustment needs to be performed. Typically, the cable will have to be lengthened or shortened to correct the problem. A reliable service resource should be used to find the location of the adjustment point for each vehicle that is worked on.

4. Inspect, test, and replace HVAC vacuum system actuators (diaphragms/motors), hoses, reservoir, check valve, and restrictors.

Vacuum actuators are used in the HVAC duct box to move the mode doors, as well as the fresh/recirculated air door. These devices are typically spring-loaded to one position and move to a second position when vacuum is applied to them. Vacuum actuators are never used for the temperature door because the temperature door needs to have more than just two or three potential positions. The HVAC control head controls when the vacuum actuators receive vacuum. These devices can fail by leaking, which causes them to not move the door to the correct position when vacuum is supplied. A vacuum pump can be used to test the vacuum actuators. A good actuator should hold vacuum for at least one minute without leaking down. If the actuator fails to hold a vacuum, then the actuator is typically serviceable without removing the whole duct box.

An HVAC system that uses vacuum as the method to move the doors in the duct box will often use a vacuum reservoir and a check valve to store vacuum and to trap vacuum in certain locations to use in case of low vacuum conditions on the engine. The vacuum reservoir is typically in the engine compartment area in varied locations. The vacuum check valve is usually somewhere near the firewall. This check valve is a one-way device that traps vacuum on the HVAC side, which prevents the doors in the duct box from changing locations in cases of low engine vacuum. Examples of low engine vacuum would be when the engine is under heavy load or even if the engine dies. Under these conditions, the vacuum check valve traps the vacuum in the HVAC system so the doors do not change location.

The vacuum reservoir and check valve should both hold a vacuum from a vacuum pump for at least one minute. Either of these components can be replaced by disconnecting the vacuum hoses and fasteners, then installing onto the new component.

5. Inspect, test, adjust, repair, or replace HVAC ducts, doors, and outlets (including dual/multi-zone systems).

The HVAC duct box contains several doors that route the airflow in the box through different passages. The position of these doors determines where the air comes from (outside or inside the cab), the temperature of the air, as well as the location in which the air is distributed. There are several methods by which the doors are moved, including with cables, linkages, vacuum actuators, and electric motors.

Dual/multi-zone HVAC systems allow the driver, passenger, and even rear seat occupants to choose the temperature of the air that blows in their areas. These types of HVAC systems are always controlled by logic devices that look for inputs from the various locations. The air box for the driver and passenger is more complex due to the need for a blend door for both the driver and passenger. Calibration of these systems will need to take place after any component is replaced or disconnected for other service procedures. There are typically two ways to calibrate these systems. They can usually be calibrated by entering manual self-diagnostic mode, which is done by pressing two of the control head buttons in a particular sequence. A scan tool is also capable of calibrating the HVAC doors.

3. Automatic and Semi-Automatic Heating, Ventilating, and A/C Systems

1. Diagnose temperature control system problems; determine needed repairs (including dual/multi-zone systems).

Automatic and semiautomatic HVAC systems utilize a logic device that automatically controls the temperature and the blower speed in the passenger compartment. This computer performs this function by monitoring many sensors, such as those listed here:

- Cabin air temperature sensor
- Ambient air temperature sensor
- Sun/light sensor
- Digital control head
- Coolant temperature sensor
- Temperature door position sensor
- Mode door position sensor
- Fresh/recirculated door position sensor

Using data from these sensors, the HVAC computer logically sends output commands to the HVAC doors, the blower motor, the A/C clutch, and the ECM to control the temperature of the passenger compartment. This computer has diagnostic capability that can store codes, as well as display them when certain buttons are pushed on the control head. Scan tools can also be used to retrieve diagnostic trouble codes from the HVAC system.

Many times during the diagnostic process of an automatic temperature control (ATC) problem, it is necessary to connect a scan tool to the vehicle to view live data from the many sensors that the system uses. It is advisable to view this data often on various vehicles in order to determine what normal values are. Knowing what the sensor values should be when they are in normal range will be helpful when troubleshooting a faulty sensor.

Scan tools are also very powerful diagnostic tools when used in the functional/output testing mode. Factory scan tools typically have the most output test functions, but many generic scanners will include some available output tests. This ability to command a device to turn on with the scan tool provides much diagnostic power. If the device will turn on with the scan tool, then the technician knows that the power, ground, fuse, relay, and wires are capable of operating; the technician likely needs to look for a faulty input problem. See Task C.2.5 for more information on dual/multi-zone systems.

2. Diagnose blower system problems; determine needed repairs (including dual/multi-zone systems).

The blower motor is usually located in the heater housing assembly. It ensures that air is circulated through the system. Blower speed on semi-automatic and automatic temperature systems is typically controlled by a logic device and a blower speed control module. The logic device (BCM, HVAC control module) looks at the various inputs from the sensors as well as the temperature at which the control head is set. All of this data is processed, and then the logic device sends a command to the blower speed control module. The blower speed control module then supplies the needed voltage to the blower motor. As the temperature moves closer to the desired setting, the system will send signals to slow the blower speed down. Vehicles with multi-zone capabilities will have a separate blower motor located in a remote location that is independently controlled. This extra blower motor allows for improved performance of the HVAC system.

3. Diagnose air distribution system problems; determine needed repairs (including dual/multi-zone systems).

Semi-automatic and fully automatic temperature control systems are controlled by one or more logic devices. The best tool for diagnosing problems on these systems is the scan tool. The scan tool allows you to see live data and trouble codes as well switch inputs coming into the HVAC logic device.

If any voltage or resistance checks need to be made, it is recommended to use a high impedance voltmeter so damage to the logic devices can be avoided. Never use analog meters or test lights on circuits controlled by solid state devices.

4. Diagnose compressor clutch control system; determine needed repairs.

The compressor clutch on a semi-automatic or automatic temperature control system operates in the same way as on a manual climate control system. Power to the clutch is supplied through a relay. The relay is controlled by the PCM, which receives signals from the various pressure and temperature sensors or over the data bus network from the HVAC or body computer. After verifying that the system is not empty, you should connect a scan tool to search for trouble codes as well as view HVAC-related data.

5. Inspect, test, adjust or replace climate and blower control sensors.

The following is a partial list of the sensors that are used in an ATC heating and cooling system, as discussed in Task C.3.1:

- Cabin air temperature sensor
- Ambient air temperature sensor
- Sun/light sensor
- Coolant temperature sensor
- Temperature door position sensor
- Mode door position sensor
- Fresh/recirculated door position sensor

The preceding temperature sensors are simply thermistors that vary their resistance as their temperature changes. Most thermistors used in automotive applications are of the negative temperature coefficient (NTC) design. This means that the resistance will increase as the temperature decreases and that the resistance will decrease as the temperature increases. These devices can get out of calibration at times and the technician will need to view the data on the scan tool to check to see if they are reading accurately. Faults in these sensors will generate trouble codes that will assist the technician in the troubleshooting process.

Potentiometers are the most common type of position sensors that are used for door position feedback. Potentiometers use three wires and are always used in conjunction with a logic device. A scan tool is the preferred tool to use when troubleshooting these sensors. Using a scan tool, the technician can view live data to watch the sensor signal voltage as well as retrieve trouble codes that may have been created by a faulty position sensor.

Photo/light sensors are typically variable resistors that change resistance as the light level changes. Faults in these sensors can be diagnosed with a scan tool as well.

6. Inspect, test, adjust, and replace door actuator(s).

The temperature blend door is controlled by an electric door actuator that is controlled by an HVAC computer. Most of these actuators have a potentiometer that is used to provide feedback to the HVAC computer about where the door is positioned. If this potentiometer fails, then the temperature door will not operate correctly.

If the temperature door actuator requires replacement, you should follow the recommended procedure. These actuators can usually be accessed without completely disassembling the HVAC duct box. The door and linkage should be checked for problems such as being stuck or bound up as the actuator is being replaced. If the door is defective, then the whole HVAC duct box will have to be removed from the vehicle.

The blend door will need to be re-calibrated after being replaced, or after the battery has been replaced on some vehicles. This process can be done manually by putting the HVAC system in diagnostic mode (by pressing the HVAC controls in a particular sequence). A calibration can also be performed using a scan tool.

7. Inspect, test, and replace heater coolant control valve and controls.

Heater control valves are usually mechanically, electrically, or vacuum controlled. When the maximum A/C mode is selected, the valve is usually closed to maximize the cooling effect. Not all blend-air systems use heater control valves because the blend door is the main device that controls the air temperature by routing the air through or around the heater core.

If the coolant control valve is defective, the hose between it and the heater core will not be hot. The indication is that hot coolant is passing the control valve but not exiting the heater core. Therefore, the heater core may be clogged and should be cleaned or replaced.

8. Inspect, test, and replace electric and vacuum motors, solenoids, and switches.

The process of testing vacuum and electric motors in an ATC system is similar to testing these devices in manual HVAC systems. One exception is that some ATC systems allow the technician to perform diagnostics using the digital control head. A series of buttons can be depressed, which causes the control head to enter diagnostic mode. In this mode, the HVAC system will sometimes display codes and data as well as perform calibration procedures on the electric actuators. Scan tools can also be used to troubleshoot problems in the HVAC solenoids and switches.

Vacuum actuators are rarely used on late-model ATC systems. However, if a technician needs to diagnose a vacuum actuator or other vacuum device, then method is the same as with a manual HVAC system. See Tasks C.2.1 and C.2.4 for further details on these steps.

9. Inspect, test, replace, and/or program Automatic Temperature Control (ATC) control panel and/or climate control computer/module; program, code, or initialize as required.

The ATC system is a very intricate system that uses a computer to perform the high-level functions. The computer can be called various names such as the HVAC controller, ECU, ECM, and BCM, as well as other possible names. The name of the device makes little difference because most computers work very much the same way. Computers need inputs, outputs, and the ability to perform logical functions very quickly. Most ATC systems give the technician two ways to retrieve trouble codes from the system. One way is to manipulate the buttons on the control head to get the system into diagnostic mode. The codes are displayed on the digital readout display of the control head. The second way to retrieve codes is to use a scan tool to interface with the computer. After the code has been retrieved, the technician should use the available service information to troubleshoot the code. Typically, a very logical troubleshooting sequence leads the technician to the correct diagnosis.

If the HVAC controller has to be replaced, it is advisable to disconnect the battery prior to beginning the process to prevent possible damage to the electrical and electronic systems. It is sometimes necessary to use a scan tool to reprogram/initialize the HVAC controller after replacement. Extra care should be taken during this process to maintain vehicle battery voltage. Many technicians will use a battery maintainer during this reprogramming process in order to limit low voltage problems.

10. Check and adjust calibration of Automatic Temperature Control (ATC) system.

Calibration of these systems will need to take place after any component is replaced or disconnected for other service procedures. There are typically two ways to calibrate these systems. They can usually be calibrated by entering manual self diagnostic mode, which is done by pressing two of the control head buttons in a particular sequence. A scan tool is also capable of calibrating the HVAC doors.

11. Diagnose data communication issues, including diagnostic trouble codes (DTCs) that affect A/C system operation.

Data communication on a late-model vehicle is vital for correct operation of any of the electronically controlled systems. All ATC systems use one or more computers to perform the function of handling the temperature and humidity inside the passenger compartment. The computers used by the ATC system must constantly communicate with each other as well as the other computers on the vehicle, such as the engine computer and the body computer. If this data communication system does not work correctly, then the ATC system, in addition to many of the other systems, will likely have problems performing normal tasks. The diagnostic system will usually set a DTC when the communication system has a fault. When this happens, the technician will need to find the correct diagnostic repair data from a database and follow the diagnostic routine to repair the problem.

The data communication network is basically made up of one, two, or three wires that form a data highway for the various computers to communicate. Problems that can occur on this network include opens, shorts, grounds, and faulty modules. The primary tool a technician uses when troubleshooting electronically controlled systems is the scan tool. When the data network is working normally, the technician has great diagnostic capability to troubleshoot problems. Viewing live data and trouble codes and commanding system output tests with the scan tool are simple tasks. However, when the data network is not functioning properly, the scan tool becomes less valuable. An inoperative data network will keep the scan tool from operating on the vehicle. When this happens, the technician must find out what is causing the failure. A digital multimeter (DMM) is the tool of choice at this stage of the process. The technician should check for the correct voltage levels, as well as for opens and grounded data bus wires. It is important to make sure the DMM is a high quality (high impedance) design in order to be safe when making voltage checks. Another tool that is often helpful when troubleshooting a data network problem is an oscilloscope. This tool is used to view the voltage signals being transmitted on the data network.

If a wire repair is needed, the technician must use high-quality wire repair methods such as soldering or crimp-and-seal connectors. Some manufacturers will require that the whole wiring harness be replaced instead of performing wire repair on the data network.

SECTION

5 Answer Keys and Explanations

INTRODUCTION

Included in this section are the answer keys for each preparation exam, followed by individual, detailed answer explanations and a reference identifying the designated task area being assessed by each specific question. This additional reference information may prove useful if you need to refer back to the task list located in Section 4 of this book for additional support.

PREPARATION EXAM 1

1. An airbag system needs to be disarmed during an evaporator core replacement. Technician A says that it is necessary to remove power from the system prior to component removal. Technician B says that the inflatable devices should be laid "face up" in a secure area after being removed from a vehicle. Who is correct?

   A. A only
   B. B only
   C. Both A and B
   D. Neither A nor B

2. An orifice tube performs any of these jobs in the A/C system EXCEPT:

   A. Meters refrigerant into the evaporator core
   B. Filters foreign particles
   C. Separates the high side from the low side
   D. Opens and closes to regulate pressure on the low side

3. A technician connects a scan tool to a late-model vehicle to troubleshoot a problem in the ATC system. The scan tool gives a fault of "data bus short to ground." Technician A says that the problem could be a data wire rubbing against a sharp body component. Technician B says that the problem could be a broken data wire. Who is correct?

   A. A only
   B. B only
   C. Both A and B
   D. Neither A nor B

4. Any of these methods of R134a A/C system leak detection are acceptable EXCEPT:
   A. Dye and black light
   B. Electronic leak detector
   C. Nitrogen and soapy water
   D. Propane burner

5. Technician A says that an electronic blend door motor uses an analog signal to open and close. Technician B says that an electronic mode door motor uses a feedback device to indicate the position of the door. Who is right?
   A. A only
   B. B only
   C. Both A and B
   D. Neither A nor B

6. What is the most likely cause for a serpentine drive belt to slip under heavy loads?
   A. Surface cracks on the inside of the belt
   B. Belt is stretched one-quarter inch
   C. Seized tensioner
   D. Glazed belt

7. Any of these practices should be followed when interacting with airbag system components EXCEPT:
   A. Unhook the negative battery cable prior to beginning work.
   B. Store inflatable components "face up" in a secure area.
   C. Disconnect the airbag indicator prior to beginning work.
   D. Carefully carry inflatable components holding them away from the body.

8. Any of these are methods to verify that the evaporator core is leaking EXCEPT:
   A. Using an electronic leak detector at the A/C duct outlets
   B. Using an ultrasonic leak detector at the HVAC case drain tube
   C. Injecting dye into the system and using a black light to inspect the substance exiting the HVAC case drain tube
   D. Using a refrigerant identifier at the low-side fitting

9. A vehicle with rear heat and air has a rear heater that does not get warm enough. Technician A says that a crimped heater pipe could be the cause. Technician B says that a partially blocked rear heater core could be the cause. Who is correct?
   A. A only
   B. B only
   C. Both A and B
   D. Neither A nor B

10. An R-134a A/C system is being recharged. Technician A says that an A/C system can be accurately charged by monitoring pressure and temperature. Technician B says that the A/C system should be treated with refrigerant dye any time that work has been performed so that future leak testing will be possible. Who is correct?

    A. A only
    B. B only
    C. Both A and B
    D. Neither A nor B

11. The HVAC fuse blows each time the A/C switch is turned on. Technician A says that an open low-side switch could be the cause. Technician B says that a shorted A/C clutch relay coil could be the cause. Who is correct?

    A. A only
    B. B only
    C. Both A and B
    D. Neither A nor B

12. A vehicle with an automatic temperature control (ATC) system is being diagnosed. Technician A says that the ATC trouble codes can be retrieved by depressing a sequence of buttons on the ATC control head. Technician B says that the trouble code reveals which area needs to be repaired in an ATC system. Who is correct?

    A. A only
    B. B only
    C. Both A and B
    D. Neither A nor B

13. Any of these conditions can cause elevated high-side pressures EXCEPT:

    A. Refrigerant overcharge
    B. Restricted airflow to the condenser
    C. Poor airflow across the evaporator
    D. A slipping fan clutch

14. Which of these oils is the standard oil used by OEMs in (nonhybrid) R-134a systems?

    A. Mineral oil
    B. 10W30
    C. Ester oil
    D. Polyalkylene glycol (PAG)

15. The heating system is being diagnosed for poor heating performance. Technician A says the problem could be a stuck open engine thermostat. Technician B says the problem could be an inoperative cooling fan. Who is right?

    A. A only
    B. B only
    C. Both A and B
    D. Neither A nor B

16. An A/C system is being diagnosed for a leak. Technician A carefully checks all system connections and says the presence of oil around the fitting of an air conditioning line or hose is an indication of a possible refrigerant leak. Technician B uses a handheld electronic leak detector and moves the hose slowly along all components of the refrigerant system. Who is correct?

    A. A only
    B. B only
    C. Both A and B
    D. Neither A nor B

17. The fins and air passages of an evaporator are heavily clogged and the airflow has been greatly reduced. Technician A says the evaporator core must be removed from the case for proper cleaning. Technician B says this condition can cause extra load on the blower motor. Who is correct?

    A. A only
    B. B only
    C. Both A and B
    D. Neither A nor B

18. Technician A says a restricted orifice tube could be caused by debris from a failing compressor. Technician B says an orifice tube restriction would cause a low-side pressure that is considerably higher than specified. Who is correct?

    A. A only
    B. B only
    C. Both A and B
    D. Neither A nor B

19. Technician A says the desiccant in a receiver/drier absorbs moisture and it can become quickly contaminated if exposed to atmosphere. Technician B says orifice tube systems use an accumulator to store excess refrigerant and to filter and dry the refrigerant. Who is correct?

    A. A only
    B. B only
    C. Both A and B
    D. Neither A nor B

20. Any of these statements about refrigerant lines are true EXCEPT:

    A. Suction lines are located between the outlet side of the evaporator and the inlet side or suction side of the compressor.
    B. Suction lines carry the low-pressure, low-temperature refrigerant vapor to the compressor where it again is moved through the system.
    C. Suction lines can be distinguished from the discharge lines by touch: They are hot to the touch when the system is operating.
    D. The suction line is larger in diameter than the liquid line because refrigerant in a vapor state takes up more room than refrigerant in a liquid state.

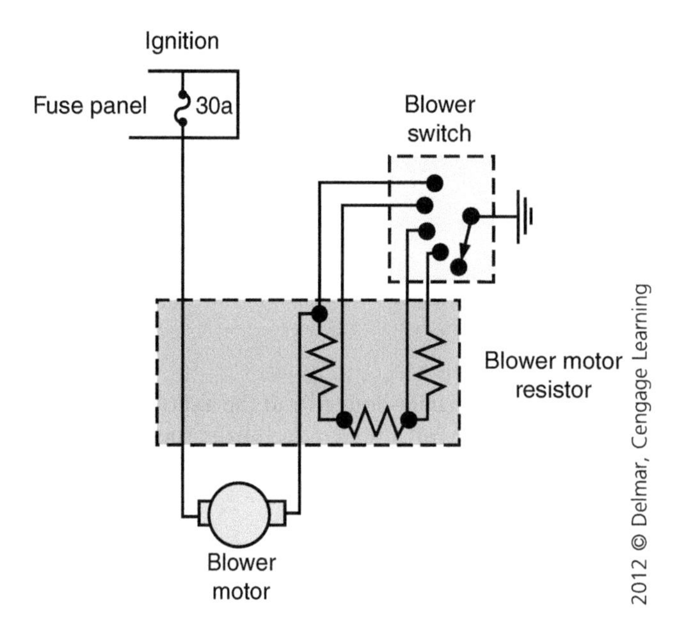

21. In the figure above, the blower motor only works on high speed. Technician A says that a faulty ground at the blower switch could be the cause. Technician B says that an open blower resistor could be the cause. Who is correct?

    A. A only
    B. B only
    C. Both A and B
    D. Neither A nor B

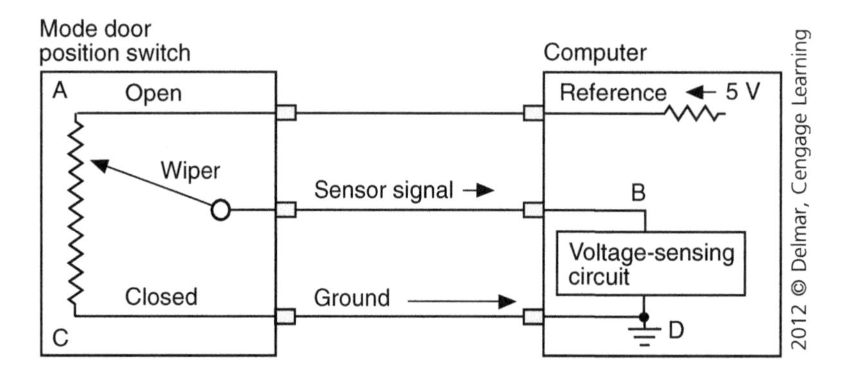

22. Referring to the figure above, the mode door actuator is inoperative. Technician A says that the computer senses the position of the mode door by monitoring the reference voltage. Technician B says that the computer provides the ground for the mode door position switch. Who is correct?

    A. A only
    B. B only
    C. Both A and B
    D. Neither A nor B

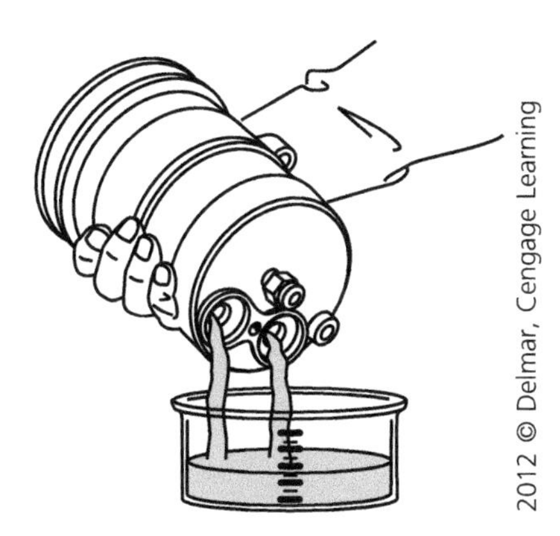

23. Referring to the figure above, Technician A says that the old compressor oil should be drained and measured when replacing an A/C compressor. Technician B says that the old oil should be added to the new compressor. Who is correct?

A. A only
B. B only
C. Both A and B
D. Neither A nor B

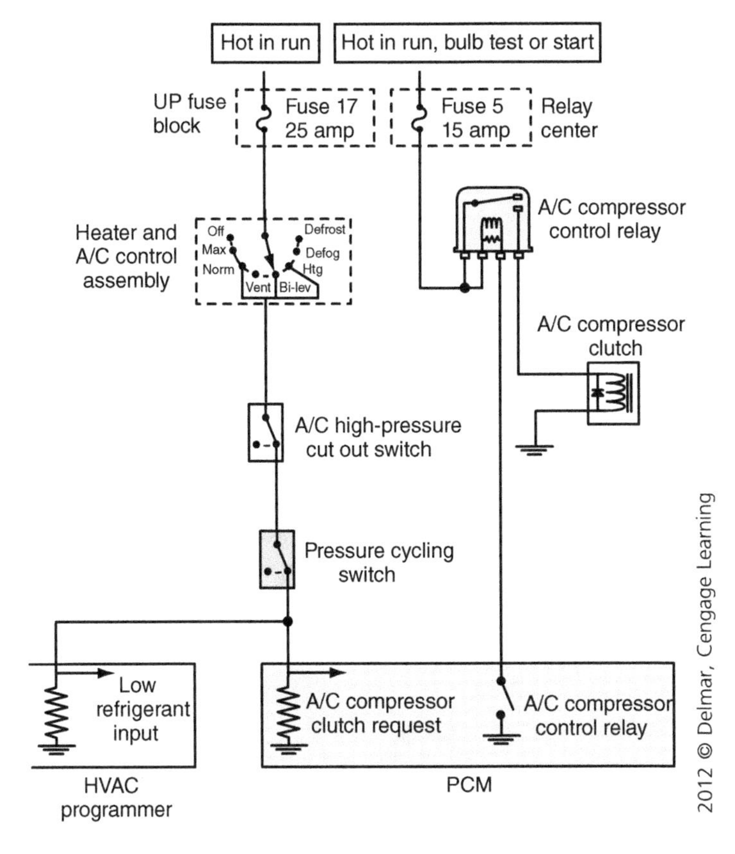

24. The A/C clutch in the figure above will not engage. Which of these would be the LEAST LIKELY cause?

A. A blown fuse 17
B. A stuck closed A/C compressor control relay
C. A faulty PCM relay driver
D. A faulty pressure cycling switch

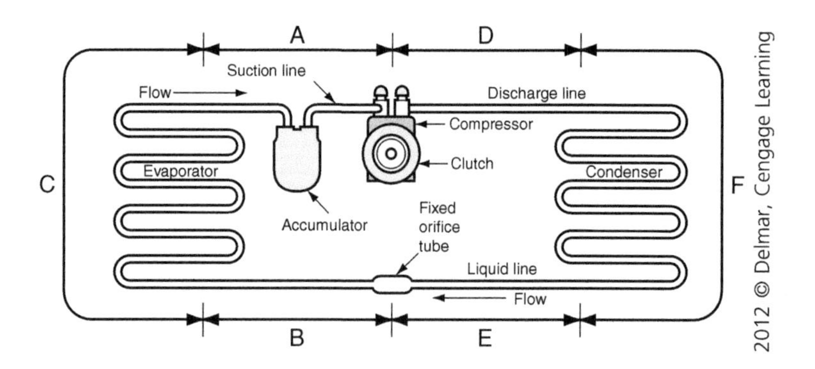

25. Referring to the figure above, list the correct flow of refrigerant from the compressor exit point back to the compressor inlet point.

A. A to C to B to E to F to D
B. D to A to C to B to E to F
C. B to C to A to D to F to E
D. D to F to E to B to C to A

26. A vehicle with a faulty A/C pressure sensor is being diagnosed. Technician A says that this device is used to provide A/C pressure feedback to a control module. Technician B says that this device is typically mounted in the low side of the A/C system. Who is correct?

A. A only
B. B only
C. Both A and B
D. Neither A nor B

27. The most likely problem that would result from an open blower resistor would be:

A. The blower would work on all lower speeds but not on high speed.
B. The blower would not work on any speed.
C. The blower would only work on high speed.
D. The blower fuse would blow each time the blower is turned on at any speed.

28. An A/C compressor cycles off when the high-side pressure builds. Technician A says that the high-pressure cutoff switch is opening the circuit to the compressor clutch. Technician B says that the pressure release valve is opening the circuit to the compressor clutch. Who is correct?

A. A only
B. B only
C. Both A and B
D. Neither A nor B

29. A performance test is being performed on the A/C system. Technician A says that the A/C system should be turned on normal A/C and the blower set to low speed. Technician B says that the engine should be at idle during this test. Who is correct?

A. A only
B. B only
C. Both A and B
D. Neither A nor B

30. The A/C compressor will not engage when the A/C is turned on. The static refrigerant pressure is 75 psi and the outside air temperature is 72°F. Technician A says that a poor connection at the pressure cycling switch could be the cause. Technician B says that a faulty A/C clutch coil could be the cause. Who is correct?

    A. A only
    B. B only
    C. Both A and B
    D. Neither A nor B

31. The vacuum supply hose has been broken at the vacuum check valve. Technician A says that the temperature door will be stuck in the full hot position. Technician B says the defrost door will be stuck in the defrost position. Who is correct?

    A. A only
    B. B only
    C. Both A and B
    D. Neither A nor B

32. A vehicle is in the shop with the complaint of the heater not being hot enough. The coolant level has been checked and is found to be full. Technician A says that the thermostat should be tested for proper operation. Technician B says that the heater control valve could be stuck open. Who is correct?

    A. A only
    B. B only
    C. Both A and B
    D. Neither A nor B

33. A vehicle is in the shop for an overheating problem. During the inspection the technician finds that the upper radiator hose collapses after the engine cools down but moves back to normal when the radiator cap is removed. The most likely cause for this problem is:

    A. A faulty water pump
    B. A malfunctioning cooling fan
    C. A faulty radiator cap
    D. A faulty thermostat

34. What function does the A/C clutch coil diode perform for the A/C system?

    A. Prevents the voltage spike from damaging other components
    B. Assists the clutch coil in creating magnetism
    C. Limits current flow to prevent the A/C fuse from overheating
    D. Connects the clutch coil to ground

35. Any of these are acceptable results from an A/C performance test EXCEPT:

    A. Duct temperatures ranging from 40°F to 50°F (4.4°C – 10°C)
    B. High-side pressure at approximately two to three times the ambient temperature
    C. Low-side pressure at approximately 50 to 70 psi
    D. Suction line is cold to the touch

36. Which device is used as the input to cause the A/C clutch to disengage under heavy engine loads?
    A. Coolant temperature sensor
    B. Throttle position sensor
    C. Oxygen sensor
    D. Power steering switch

37. Vacuum actuators can be used for any of these duct box doors EXCEPT:
    A. Temperature door
    B. Defrost mode door
    C. Fresh air door
    D. Heat mode door

38. The thermal limiter is being diagnosed on a late-model vehicle. Technician A says that this device can be checked with a scan tool by using a function test of the cooling fan. Technician B says that this device can be checked with an ohmmeter when the power is turned off. Who is correct?
    A. A only
    B. B only
    C. Both A and B
    D. Neither A nor B

39. Technician A says that the evaporator temperature switch is a thermostatic device that opens when the evaporator core gets near the freezing level. Technician B says that the evaporator temperature sensor is a thermistor that varies as the temperature changes in the evaporator core. Who is correct?
    A. A only
    B. B only
    C. Both A and B
    D. Neither A nor B

40. A vehicle with automatic temperature control (ATC) is malfunctioning. A scan tool is connected and a code is retrieved from the system concerning the cabin temperature sensor. Technician A says that the sensor is located near the engine air filter. Technician B says the sensor can be checked with an ohmmeter to see if it has the correct resistance. Who is correct?
    A. A only
    B. B only
    C. Both A and B
    D. Neither A nor B

41. Any of these methods are used to move the temperature door EXCEPT:
    A. Electronic actuators
    B. Vacuum actuators
    C. Cables
    D. Linkages

42. An A/C compressor is noisy when the clutch is engaged but the noise goes away when the clutch is disengaged. Which of these is the most likely cause?
    A. Faulty compressor clutch pulley bearing
    B. Low refrigerant charge
    C. Internal compressor damage
    D. Overtightened drive belt

43. A refillable A/C refrigerant cylinder is considered full when it reaches what capacity by weight?
    A. 50 percent
    B. 60 percent
    C. 70 percent
    D. 80 percent

44. The refrigerant is being recovered from an A/C system. Five minutes after the recovery process is complete, the low-side pressure loses the vacuum and the pressure rises above zero. This condition indicates:
    A. There is still some refrigerant in the system.
    B. There is excessive oil in the refrigerant system.
    C. The refrigerant system is leaking.
    D. There is excessive moisture in the refrigerant system.

45. A cycling clutch orifice tube (CCOT) A/C system is operating at 82°F (27.8°C) ambient temperature, the compressor clutch cycles several times per minute, and the suction line is warm. The high-side gauge shows lower than normal pressures. The most likely cause of this problem could be:
    A. A low refrigerant charge
    B. A flooded evaporator
    C. A restricted accumulator
    D. An overcharge of refrigerant

46. A vehicle has a problem of a stalling engine when the steering wheel is turned to full lock. This problem happens only when the air conditioning is on. Technician A says that the vehicle may have a bad power steering pressure switch. Technician B says this could occur if the air conditioning belt is loose. Who is correct?
    A. A only
    B. B only
    C. Both A and B
    D. Neither A nor B

47. An air distribution vacuum actuator circuit is being tested using a vacuum gauge. The gauge is connected into the vacuum line going to the engine, and a zero vacuum reading is displayed. This would indicate:
    A. The HVAC vacuum switching valve is faulty.
    B. The line to the engine is plugged, disconnected, or kinked.
    C. The vacuum actuator is defective.
    D. The actuator should be operational.

48. The high-side pressure on a system with a cycling clutch and pressure cycling switch is 245 psi, the low-side pressure hovers around 28 psi, and the ambient temperature is 98°F. What do these gauge readings indicate?

    A. The system is undercharged.
    B. The system is overcharged.
    C. The system is normal.
    D. The evaporator pressure regulator is bad.

49. A vehicle is being diagnosed for the cause of a blown fuse in the blower motor circuit. Technician A says a short to ground in the circuit caused the fuse to blow. Technician B says an open field winding in the fan motor could have caused the fuse to blow. Who is correct?

    A. A only
    B. B only
    C. Both A and B
    D. Neither A nor B

50. While testing a compressor clutch, a fused jumper wire is used to bypass the load side of the A/C clutch relay. This does not cause the compressor clutch to engage. The LEAST LIKELY cause of this would be:

    A. An open compressor clutch coil
    B. A shorted compressor clutch coil
    C. An open compressor clutch coil ground circuit
    D. An open relay power feed circuit

PREPARATION EXAM 2

1. What would be the most likely result of a missing A/C compressor clutch diode?

    A. Inoperative A/C compressor
    B. Damage to the ECM
    C. Compressor will not disengage
    D. Rear defogger will not engage

2. Technician A says that the service fittings for R-134a are the same as the service fittings for R-12. Technician B says that the low-side service fitting is a 16 mm quick-connect style and the high-side fitting is a 13 mm quick-connect style. Who is correct?

    A. A only
    B. B only
    C. Both A and B
    D. Neither A nor B

3. An inoperative cable-controlled heater control valve is being diagnosed. The control moves freely, but the valve does not respond. Technician A says the cable housing clamp may be loose at the control head (panel) end. Technician B says the cable may be rusted in the housing. Who is correct?

    A. A only
    B. B only
    C. Both A and B
    D. Neither A nor B

4. Technician A says that the scan tool receives data from the vehicle by communicating on the data bus network. Technician B says that if one of the two data bus wires becomes broken, then the network can still communicate on the remaining good wire but will set a trouble code. Who is correct?

   A. A only
   B. B only
   C. Both A and B
   D. Neither A nor B

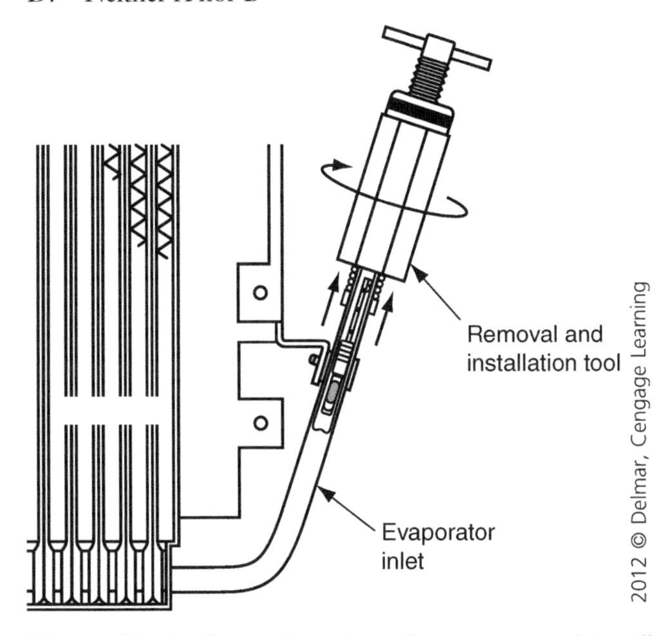

5. The tool in the figure above is used to remove and install which A/C component?

   A. The evaporator core
   B. The heater core
   C. The TXV
   D. The orifice tube

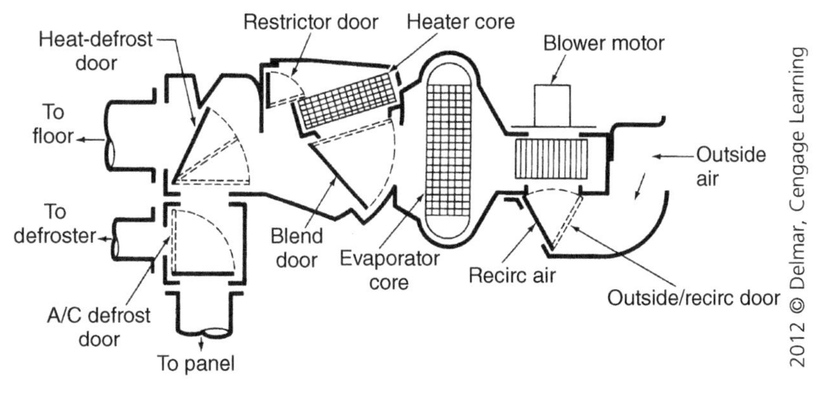

6. Referring to the figure above, the blend door gets jammed in the middle of the travel area. What would be the most likely system fault?

   A. The HVAC air would be pulled from outside the car.
   B. The HVAC air would only come out of the heat and defroster grids.
   C. The HVAC air temperature could not be regulated.
   D. The HVAC air would only come out of the vent and heat ducts.

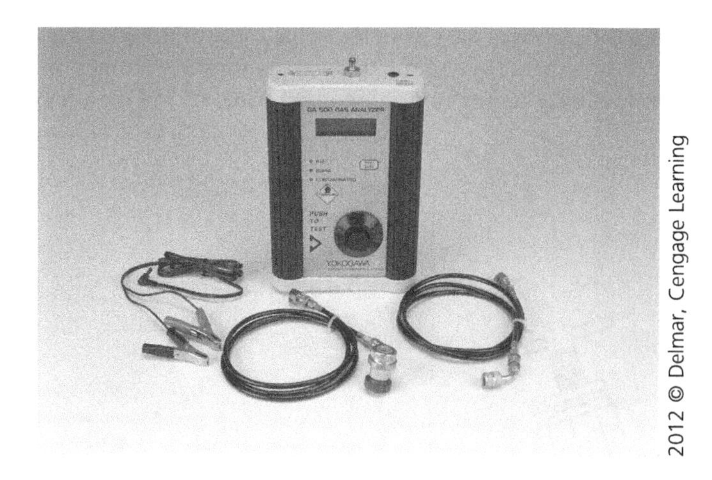

7. Which A/C-related procedure would the above tool be used for?

   A. Weighing refrigerant during recharging
   B. Testing the refrigerant for impurities
   C. Testing the A/C system performance
   D. Testing the coolant for electrolysis

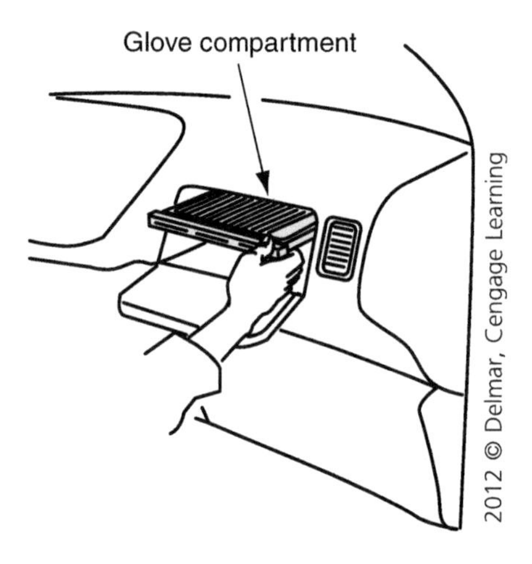

8. Referring to the figure above, what would be the HVAC malfunction if the component became restricted?

   A. AC not cold enough
   B. Heater too warm
   C. Heater not warm enough
   D. Low airflow from the ducts

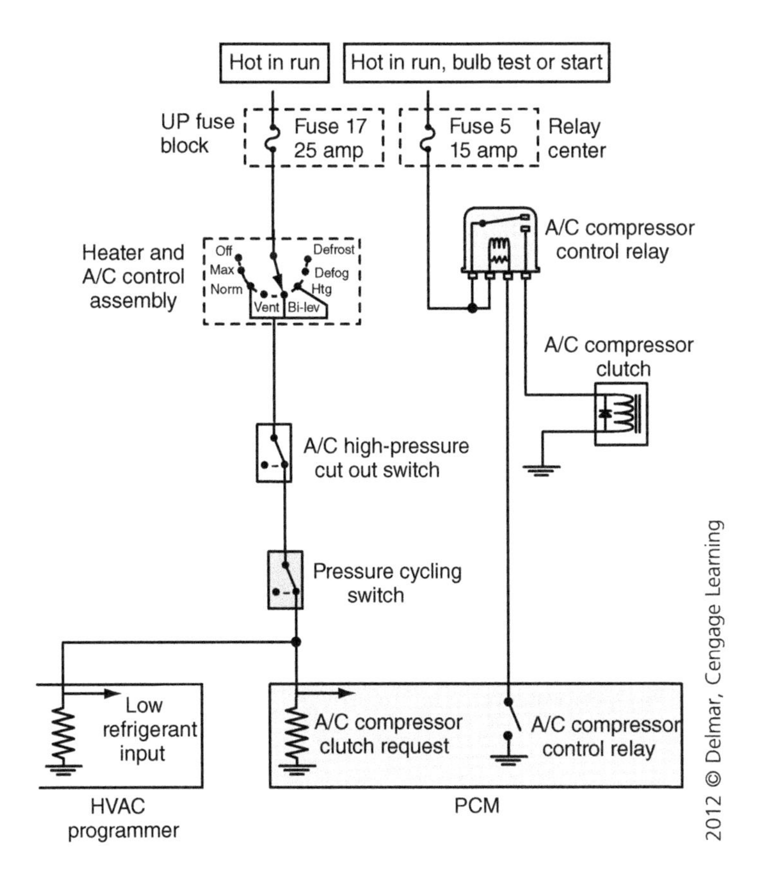

9. Referring to the figure above, the A/C compressor clutch will not engage. Technician A says that the A/C clutch relay could be bypassed with a fused jumper to test the load side of the circuit. Technician B says that fuse 17 sends power to the A/C compressor control relay. Who is correct?

A. A only
B. B only
C. Both A and B
D. Neither A nor B.

10. A hissing noise is heard in the engine compartment for approximately one minute after the engine is shut off. Technician A says that the A/C system equalization makes an audible noise when the engine is shut off. Technician B says the heater core makes an audible noise when the engine is shut off. Who is correct?

A. A only
B. B only
C. Both A and B
D. Neither A nor B

11. The vacuum check valve was missing from the HVAC vacuum circuit but the remaining hoses were still connected. Technician A says that the modes would be stuck in the heat mode and not move in any other positions. Technician B says that the mode actuators would work normally until the vehicle drove up steep hills. Who is correct?

A. A only
B. B only
C. Both A and B
D. Neither A nor B

12. A performance test of the A/C system is being performed on a late-model vehicle. Technician A says that a thermometer should be installed into the center duct during the test. Technician B says that the A/C refrigerant should be recovered during the test. Who is correct?

    A. A only
    B. B only
    C. Both A and B
    D. Neither A nor B

13. Technician A says a restricted orifice tube will cause elevated pressures on the low and high sides of the A/C system. Technician B says that faulty reed valves in the compressor can cause elevated pressures on the low and high side of the A/C system. Who is correct?

    A. A only
    B. B only
    C. Both A and B
    D. Neither A nor B

14. Technician A says that an overheating engine can cause the A/C to shut down and blow warm air. Technician B says that running the vehicle at wide-open throttle (WOT) for several miles can cause the A/C to lose performance. Who is correct?

    A. A only
    B. B only
    C. Both A and B
    D. Neither A nor B

15. A vehicle with an R-134a A/C system is being leak tested. Which of these methods is the LEAST LIKELY method to find a refrigerant leak?

    A. Visual inspection
    B. Electronic detector
    C. Nitrogen injection
    D. Propane injection

16. An A/C system that has been worked on at another shop is being diagnosed. It is suspected that a refrigerant containing "stop leak" has been added to the system. Technician A says that the refrigerant identifier will detect this substance. Technician B says that this substance can damage A/C recovery/recharging machines. Who is correct?

    A. A only
    B. B only
    C. Both A and B
    D. Neither A nor B

17. A pressure switch needs to be replaced on a late-model vehicle. Technician A says that the A/C refrigerant does not have to be recovered if the switch has a Schrader valve under it. Technician B says that the refrigerant does need to be recovered if the switch is retained by a snap ring. Who is correct?

    A. A only
    B. B only
    C. Both A and B
    D. Neither A nor B

18. A vehicle with automatic temperature control (ATC) has a "no bus" message on the control head. Technician A says that the A/C fuse is likely blown and is causing this message. Technician B says that the HVAC system may move to defrost position with heated air from the duct. Who is correct?

    A. A only
    B. B only
    C. Both A and B
    D. Neither A nor B

19. Any of these items need to be checked and inspected when replacing an A/C compressor EXCEPT:

    A. The piston ring end clearance
    B. The clutch air gap
    C. The number of pulley grooves
    D. The location of the mounting holes

20. Which tool should be used to test for problems in a vacuum-operated HVAC system?

    A. Vacuum pump
    B. Pressure gauge
    C. Pressure pump
    D. Vacuum gauge

21. A knocking noise is heard in the compressor area that is audible when the compressor is engaged but it goes away when the compressor turns off. Technician A says that loose compressor mounting bolts could be the cause. Technician B says that a discharge line rubbing a compressor mounting bracket could be the cause. Who is correct?

    A. A only
    B. B only
    C. Both A and B
    D. Neither A nor B

22. The receiver/drier is being replaced on a late-model vehicle. Technician A says that the drier should be installed last to prevent it from being exposed to moisture for a prolonged time. Technician B says that some refrigerant oil should be added to the drier prior to installation. Who is correct?

    A. A only
    B. B only
    C. Both A and B
    D. Neither A nor B

23. The windshield fogs up when the defroster is turned on and the cab is filled with a sweet smell. Which of these is the most likely cause?

    A. Blown head gasket
    B. Leaking heater control valve
    C. Leaking evaporator core
    D. Leaking heater core

24. Technician A says that electronic mode actuators can be recalibrated by using a scan tool. Technician B says that electronic blend actuators can be recalibrated by using a scan tool. Who is correct?

    A. A only
    B. B only
    C. Both A and B
    D. Neither A nor B

25. A vehicle's cooling system is being inspected during a 30,000-mile service. Technician A says that the radiator cap should be pressure tested to check for correct operation. Technician B says that the coolant freeze protection should be checked with a voltmeter. Who is correct?

    A. A only
    B. B only
    C. Both A and B
    D. Neither A nor B

26. What is the most likely cause for a blower motor fuse to blow?

    A. Blower resistor open
    B. Faulty blower motor ground circuit
    C. Blower motor is seized
    D. Burned blower relay contact

27. Technician A says that a restricted cabin air filter can cause reduced airflow at all of the duct outlet locations. Technician B says that the cabin air filter should be inspected during routine preventative maintenance activities on a vehicle. Who is correct?

    A. A only
    B. B only
    C. Both A and B
    D. Neither A nor B

28. All of the vacuum controls are inoperative with the engine running at idle speed. Technician A says the problem could be caused by a disconnected vacuum supply hose. Technician B says the manifold vacuum fitting may be blocked. Who is correct?

    A. A only
    B. B only
    C. Both A and B
    D. Neither A nor B

29. What is the LEAST LIKELY cause for a heater failing to produce adequate hot air?

    A. Mode door out of calibration
    B. Heater core with partially blocked passages
    C. Heater control valve misadjusted
    D. Blend air door misadjusted

30. A vehicle with automatic temperature control (ATC) has a problem with the A/C system. The set temperature and the actual temperature in the cab differ by 20 degrees after 15 minutes of operation. Technician A says the cabin air temperature sensor may be defective. Technician B says the temperature blend door may be sticking. Who is correct?

    A. A only
    B. B only
    C. Both A and B
    D. Neither A nor B

31. A vehicle has the temperature selector set to the cool position but the air from the vents is warm. The most likely problem is:

    A. The engine cooling fan is inoperative.
    B. The engine coolant level is low.
    C. The heater core's coolant passages are restricted.
    D. An air blend door is stuck.

32. Technician A says that stored refrigerant should be kept warm by providing a heat source near the storage area. Technician B says that recovered refrigerant should be kept in a DOT 39 cylinder. Who is correct?

    A. A only
    B. B only
    C. Both A and B
    D. Neither A nor B

33. A refrigerant identifier is connected to an A/C system and gives the reading of 95 percent R134a and 5 percent R12. Technician A says that this system can be safely recovered into the R134a recovery machine. Technician B says this system has likely been retrofitted using nonstandard procedures. Who is correct?

    A. A only
    B. B only
    C. Both A and B
    D. Neither A nor B

34. Technician A says that all A/C repair shops are required to use SAE-approved recovery equipment. Technician B says that all individuals who service A/C systems must be certified by a recognized body on how to properly handle refrigerants. Who is correct?

    A. A only
    B. B only
    C. Both A and B
    D. Neither A nor B

35. Technician A says that the suction line should be hot to the touch while the A/C system is operating. Technician B says that the liquid line should be hot to the touch while the system is operating. Who is correct?

    A. A only
    B. B only
    C. Both A and B
    D. Neither A nor B

36. The purpose of the A/C pressure sensor is to:

    A. Open and close as refrigerant pressure changes in the low side of the system
    B. Open and close as refrigerant pressure changes in the high side of the system
    C. Send a variable signal to a processor as evaporator temperature changes
    D. Send a variable signal to a processor as high-side pressure changes

37. The A/C system refrigerant charge amount needs to be determined on a late-model vehicle. Technician A says that the emissions label contains the refrigerant charge amount. Technician B says that the RPO decal contains the refrigerant charge amount. Who is correct?

    A. A only
    B. B only
    C. Both A and B
    D. Neither A nor B

38. What happens to the A/C system during the evacuation process?

    A. The refrigerant is removed from the A/C system.
    B. The A/C system is pulled into a deep vacuum to remove any moisture.
    C. The refrigerant is filtered and cleaned.
    D. The refrigerant oil is removed from the A/C system.

39. Any of these components are A/C pressure devices EXCEPT:

    A. Thermal limiter
    B. AC pressure transducer
    C. Dual pressure switch
    D. Cycling switch

40. Any of these steps should be followed when working around airbag systems EXCEPT:

    A. Store the airbag components face down on the bench when not in use.
    B. Disconnect the negative battery cable prior to beginning work.
    C. Use caution when reconnecting airbag components.
    D. Walk with the airbag components facing away from the face.

41. A vehicle A/C system is being recharged manually with a manifold set and a 30-pound cylinder of refrigerant. Technician A says that the high-side manifold valve should be open when charging with the system turned on. Technician B says that the low-side manifold valve should be closed when charging with the system turned on. Who is correct?

    A. A only
    B. B only
    C. Both A and B
    D. Neither A nor B

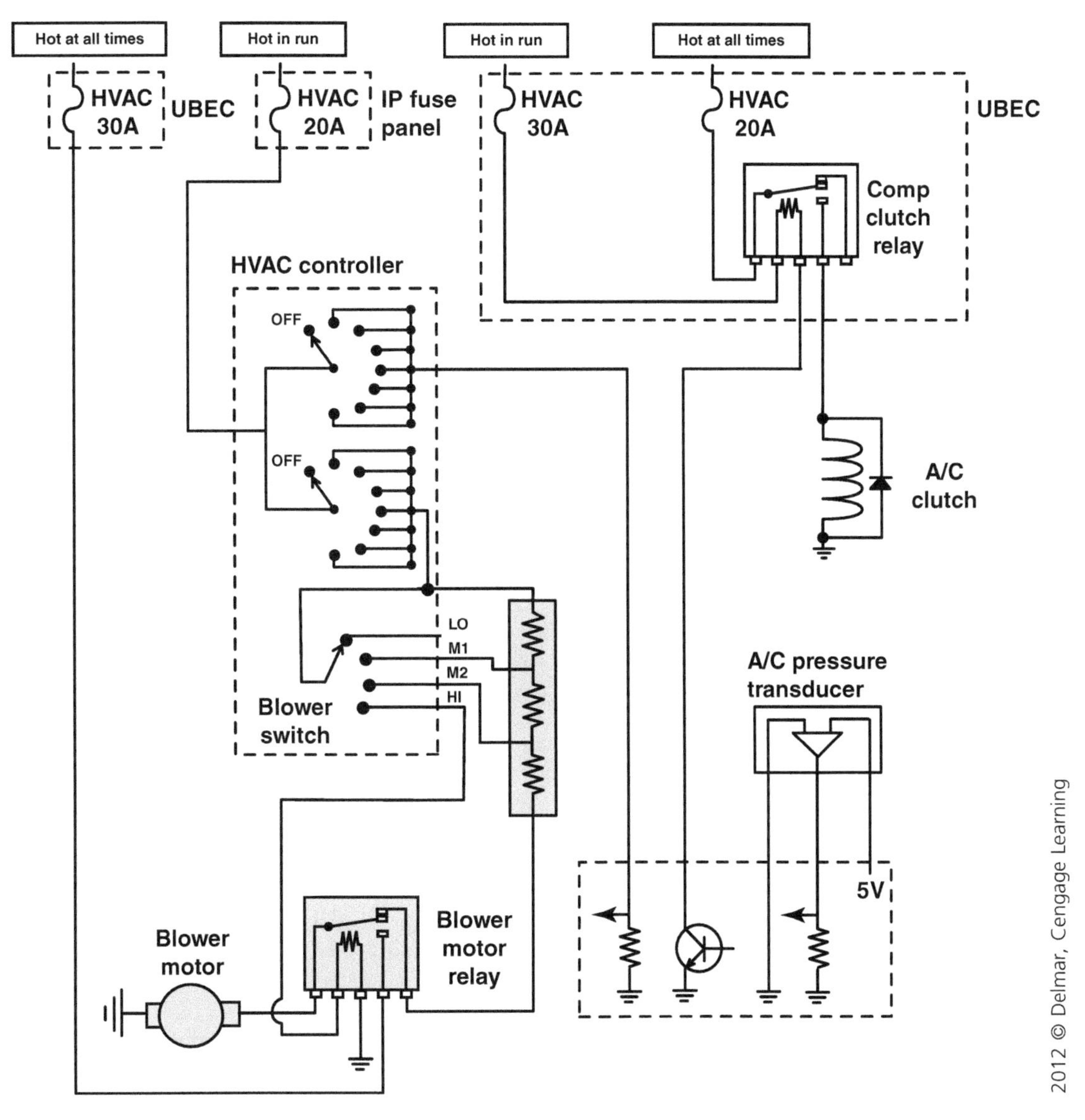

42. Referring to the figure above, the blower will not work at any time. Technician A says that the 20 amp HVAC fuse could be blown. Technician B says that the blower motor ground could be faulty. Who is correct?

A. A only
B. B only
C. Both A and B
D. Neither A nor B

43. A serpentine belt is being replaced and the tensioner will not snap back after being released. Technician A says that the tensioner spring could be broken and the tensioner will need to be replaced. Technician B says that the idler pulley is jammed and will need to be replaced. Who is correct?

A. A only
B. B only
C. Both A and B
D. Neither A nor B

44. Technician A says that R-134a systems use mineral oil to lubricate the compressor. Technician B says that R-12 systems use PAG oil to lubricate the compressor. Who is correct?

    A. A only
    B. B only
    C. Both A and B
    D. Neither A nor B

45. Technician A says that the HVAC control panel directs the signal to the mode actuators to determine where the air is discharged. Technician B says that some HVAC control panels contain logic capabilities. Who is correct?

    A. A only
    B. B only
    C. Both A and B
    D. Neither A nor B

46. A heater core is suspected to be restricted and causing the heater to not be warm enough. Technician A says that a temperature drop test should be performed on the heater inlet and outlet hoses. Technician B says that the heater system should create about 140°F air at the vents. Who is correct?

    A. A only
    B. B only
    C. Both A and B
    D. Neither A nor B

47. Any of these steps should be followed when replacing an automatic temperature control (ATC) controller EXCEPT:

    A. Disconnect the positive battery cable.
    B. Connect a ground strap to your wrist and connect to a metallic component.
    C. Use a battery maintainer while reprogramming the new controller.
    D. Disconnect the negative battery cable.

48. Technician A says that the line exiting the condenser should be hotter than the line entering the condenser. Technician B says that the suction line should be cold to the touch when the A/C system is operating. Who is correct?

    A. A only
    B. B only
    C. Both A and B
    D. Neither A nor B

49. Before replacing an HVAC electronic control panel, the technician should:

    A. Remove the control cables from the vehicle.
    B. Disconnect the negative battery cable.
    C. Disassemble the dash panel.
    D. Apply dielectric grease to the switch contacts.

50. The compressor clutch will not disengage when the A/C control is switched off. The most likely problem is:

   A. The A/C pressure cutoff switch is stuck open.
   B. The compressor clutch coil is shorted to ground.
   C. The low-pressure switch has an open wire.
   D. The compressor coil feed circuit is shorted to voltage.

PREPARATION EXAM 3

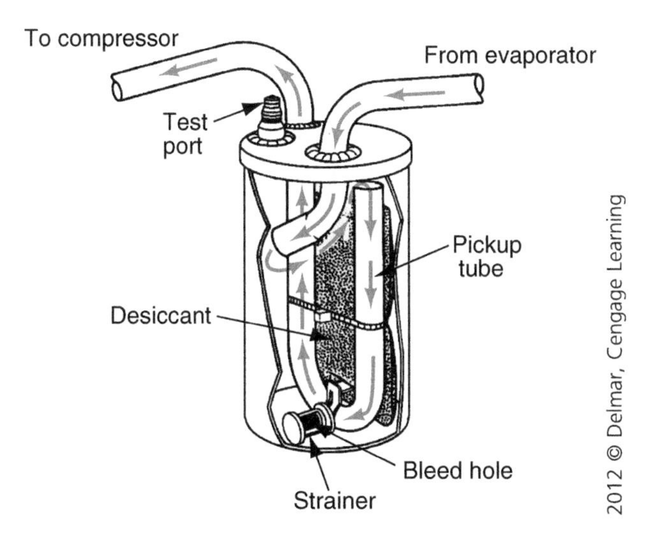

1. Which A/C system device is in the figure above?

   A. Evaporator core
   B. Receiver/drier
   C. Muffler
   D. Accumulator/drier

2. What is the LEAST LIKELY condition that could cause a blower resistor thermal element to burn up?

   A. Tight blower motor bearing
   B. Open blower motor winding
   C. Wire rubbing a metal bracket
   D. Blower motor fan blade rubbing the duct

3. An R-134a A/C system has a damaged service fitting. Technician A says the fitting needs to be mid-seated after the manifold hose adapter is installed in order to read A/C pressure. Technician B says that the Schrader valve in the fitting can be replaced without recovering the refrigerant. Who is correct?

   A. A only
   B. B only
   C. Both A and B
   D. Neither A nor B

4. A pulse width modulated electric cooling fan motor is being diagnosed. Technician A says that the HVAC control module energizes these motors by signaling a fan control driver module. Technician B says that these fans operate at two speeds. Who is correct?

   A. A only
   B. B only
   C. Both A and B
   D. Neither A nor B

5. Any of these methods are used to detect an A/C refrigerant leak on an R134a system EXCEPT:

   A. Flame-type leak detector
   B. Halogen leak detector
   C. Dye and UV light
   D. Audible noise from the leak

6. An A/C system has just been recovered with a multi-purpose recovery/recharging machine. Which of these statements is true of the condition of the A/C system?

   A. All of the oil has been removed from the system.
   B. The system is ready for normal use.
   C. All refrigerant has been removed from the system.
   D. Humidity has been removed from the refrigerant in the system.

7. A vehicle with electronic dual-zone climate control has a problem of the passenger side only blowing hot air no matter what the passenger setting is adjusted to. Technician A says that passenger blend door actuator could be defective. Technician B says that the passenger defroster mode actuator could be defective. Who is correct?

   A. A only
   B. B only
   C. Both A and B
   D. Neither A nor B

8. Technician A says that automatic temperature control (ATC) control heads are typically a digital design. Technician B says that some automatic temperature control (ATC) systems use a cable-operated heat mode actuator. Who is correct?

   A. A only
   B. B only
   C. Both A and B
   D. Neither A nor B

9. A vehicle is being evacuated after an A/C repair but the system will not drop below 10 in. Hg after 20 minutes. Technician A says that this is normal and to vacuum for 10 more minutes. Technician B says there is likely a leak in the refrigerant system. Who is correct?

   A. A only
   B. B only
   C. Both A and B
   D. Neither A nor B

10. Which of these methods is the most likely way to detect sealer in an A/C refrigerant system?

    A. Recovery machine
    B. Vacuum pump
    C. Manifold set
    D. Orifice bleed detection kit

11. Technician A says that a restricted cabin air filter will cause reduced fuel economy. Technician B says that a cabin air filter prevents foreign debris from getting on the evaporator core. Who is correct?

    A. A only
    B. B only
    C. Both A and B
    D. Neither A nor B

12. During a diagnosis of an A/C system, oil residue is found around the high-pressure relief valve and the A/C refrigerant charge has been determined to be low. Technician A says the cause could be restricted air passages through the condenser. Technician B says the refrigerant system may have been overcharged. Who is correct?

    A. A only
    B. B only
    C. Both A and B
    D. Neither A nor B

13. The air gap on a compressor clutch is found to be twice the specified amount. Technician A says this could cause an intermittent scraping noise with the engine running and the compressor clutch disengaged. Technician B says this could cause a slipping compressor clutch when the compressor is engaged. Who is correct?

    A. A only
    B. B only
    C. Both A and B
    D. Neither A nor B

14. A knocking sound is heard from the area of the A/C compressor when in operation. When the compressor is shut off, the noise stops. The A/C system cools well, and there are no indications of A/C system problems. Technician A says the noise could be caused by a broken compressor mounting bracket. Technician B says that the noise could be caused by loose fasteners of the compressor mounting. Who is correct?

    A. A only
    B. B only
    C. Both A and B
    D. Neither A nor B

15. A condenser is being diagnosed for an airflow problem. Technician A says a slipping fan clutch will reduce ram airflow. Technician B says reduced airflow across the condenser results in low suction pressure. Who is correct?

 A. A only
 B. B only
 C. Both A and B
 D. Neither A nor B

16. A vehicle is being diagnosed for an air conditioning system that cools satisfactorily during the early morning or late evening, but does not cool during the hot part of the day. During the performance, the low-side gauge starts out reading normal, then drops into a vacuum. Technician A says that ice could be forming in the expansion valve. Technician B says that the drier could be saturated with moisture. Who is correct?

 A. A only
 B. B only
 C. Both A and B
 D. Neither A nor B

17. A vehicle is being diagnosed for the problem of the compressor clutch failing to engage with the A/C turned on. A voltage measurement shows there is 12V at the clutch coil feed wire. Technician A says that the clutch coil could be bad. Technician B says that the clutch coil ground could be bad. Who is correct?

 A. A only
 B. B only
 C. Both A and B
 D. Neither A nor B

18. What is the most likely A/C problem that will occur if the capillary tube on the TXV is broken?

 A. The air conditioning will always blow at full cold.
 B. The air conditioning will not blow cold enough.
 C. It will become impossible to charge the air conditioning.
 D. It will not affect the air conditioning.

19. The blower motor fuse blows each time the blower switch is turned on. Technician A says that an open low-side switch could be the cause. Technician B says that a shorted blower motor relay could be the cause. Who is correct?

 A. A only
 B. B only
 C. Both A and B
 D. Neither A nor B

20. Technician A says that a typical A/C pressure sensor has five wires. Technician B says that a typical A/C pressure sensor operates on a 24 volt signal. Who is correct?

 A. A only
 B. B only
 C. Both A and B
 D. Neither A nor B

21. The A/C compressor has a squealing sound when the compressor engages. Technician A says that the belt is likely not adjusted to specifications. Technician B says that the air gap on the A/C compressor clutch plate may be too wide. Who is correct?

    A. A only
    B. B only
    C. Both A and B
    D. Neither A nor B

22. The inside of the windshield has a foggy film. Technician A says the engine coolant level should be checked. Technician B says the heater core may be leaking. Who is correct?

    A. A only
    B. B only
    C. Both A and B
    D. Neither A nor B

23. Which device is used as the input to cause the A/C clutch to disengage under high engine temperatures?

    A. Coolant temperature sensor
    B. Throttle position sensor
    C. Oxygen sensor
    D. Power steering switch

24. The engine overheats when the vehicle is sitting in heavy traffic. When the vehicle is driven at highway speeds, the engine operates at the normal temperature. Technician A says a broken radiator fan shroud could be the cause. Technician B says a faulty thermostatic fan clutch could be the cause. Who is correct?

    A. A only
    B. B only
    C. Both A and B
    D. Neither A nor B

25. Any of these conditions could cause the cooling system to develop a voltage potential EXCEPT:

    A. Faulty water pump
    B. A poor blower motor ground
    C. Coolant acidity is too high
    D. Loose negative battery cable connection at the engine block

26. The blower motor operates slower than normal at all speed settings. A voltage test is performed at the blower motor connector with the switch in the high-speed switch position and 7.5 volts is measured. Technician A says that the cause could be a bad blower motor. Technician B says the problem could be a high resistance at the blower motor ground. Who is correct?

    A. A only
    B. B only
    C. Both A and B
    D. Neither A nor B

27. During an A/C performance test on a TXV system that uses R-134a on a 78°F (25.6°C) day after 12 minutes of A/C operation, the pressure on the high-side gauge was 350 psi (2406.9 kPa) and the pressure on the low-side gauge was 50 psi (343.8 kPa). Technician A says that these readings are normal for the temperature and conditions. Technician B says that these readings could be caused by a refrigerant overcharge. Who is correct?

A. A only
B. B only
C. Both A and B
D. Neither A nor B

28. A vehicle with an automatic temperature control (ATC) A/C system is being diagnosed. Technician A says that some systems have to be recalibrated after the vehicle battery is replaced. Technician B says that these systems use a sun load sensor to calculate radiant heat load. Who is correct?

A. A only
B. B only
C. Both A and B
D. Neither A nor B

29. An air conditioning system equipped with an accumulator/drier is a:

A. Thermostatic expansion valve system
B. Fixed orifice tube system
C. Receiver/drier system
D. Automatic temperature control system

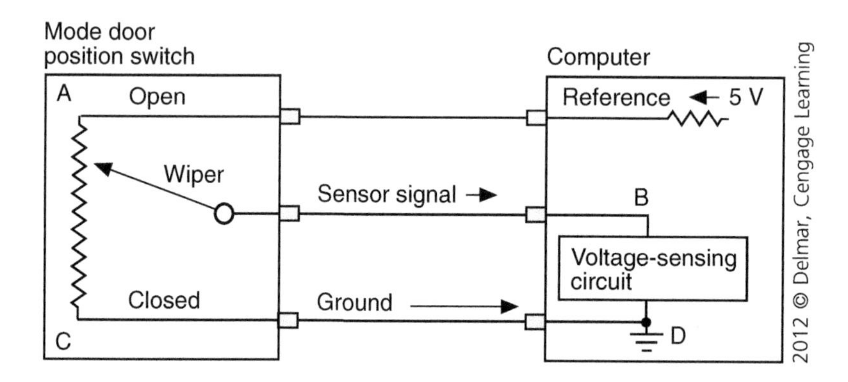

30. Referring to the figure above, the ground terminal at the computer is disconnected. Technician A says that the mode door will still operate but will set a code in the memory. Technician B says that the sensor signal will read 5 volts all of the time, no matter where the mode door is. Who is correct?

A. A only
B. B only
C. Both A and B
D. Neither A nor B

31. Technician A says that a dual-zone climate control system uses two heater cores. Technician B says that a dual-zone climate control system uses two evaporator cores. Who is correct?

    A. A only
    B. B only
    C. Both A and B
    D. Neither A nor B

32. Technician A says that high ambient temperature levels will cause the high-side pressure to increase. Technician B says that high ambient temperature levels will cause the low-side pressure to decrease. Who is correct?

    A. A only
    B. B only
    C. Both A and B
    D. Neither A nor B

33. What is the most likely cause of water dripping out of a vehicle after running the A/C system and being parked on a hot and humid day?

    A. Transmission cooler
    B. Power steering
    C. AC compressor
    D. Duct box condensate tube

34. Technician A says that an electronic HVAC control head could be damaged by static electricity. Technician B says that static electricity can be eliminated if the technician never touches any metal objects. Who is correct?

    A. A only
    B. B only
    C. Both A and B
    D. Neither A nor B

35. Which of these is LEAST LIKELY to be performed during an A/C system recovery procedure?

    A. Refrigerant is removed from a vehicle.
    B. Refrigerant is filtered by the recovery machine.
    C. Most of the A/C system oil is removed from the vehicle.
    D. Refrigerant is weighed by the recovery machine.

36. Technician A says vacuum-operated HVAC duct actuators use a spring to cause them to move to a known position when no vacuum is applied. Technician B says that cable-operated HVAC duct doors can be adjusted by shortening and lengthening the Bowden cable. Who is correct?

    A. A only
    B. B only
    C. Both A and B
    D. Neither A nor B

37. Technician A says that cable-operated HVAC duct doors produce an audible sound when the doors hit each end of travel. Technician B says that vacuum-operated HVAC duct doors need to be calibrated after the battery has been disconnected. Who is correct?

A. A only
B. B only
C. Both A and B
D. Neither A nor B

38. Technician A says that an electronic scan tool can be used to retrieve ATC diagnostic trouble codes from a vehicle. Technician B says that the ATC control head can be used to retrieve diagnostic trouble codes from the engine computer. Who is correct?

A. A only
B. B only
C. Both A and B
D. Neither A nor B

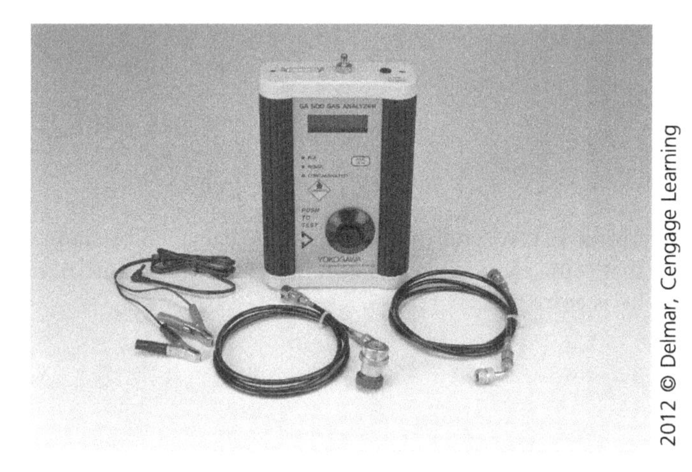

39. The tool referenced in the figure above is used:

A. To test the refrigerant for purity
B. To read A/C pressures to assist in troubleshooting for problems
C. To evacuate the A/C system to remove the moisture
D. To recover the refrigerant

40. A recovery machine shuts down during the recovery process. An indicator light is signaling FULL. Technician A says the car's system was overcharged and has reached the normal refrigerant level. Technician B says the recovery cylinder may be 80 percent full. Who is correct?

A. A only
B. B only
C. Both A and B
D. Neither A nor B

41. Frost is forming on one of the condenser tubes near the bottom of the condenser. The most likely cause of this problem would be:

A. Restricted airflow at the condenser
B. A weak A/C compressor
C. A restricted refrigerant passage in the condenser
D. A restricted fixed orifice tube

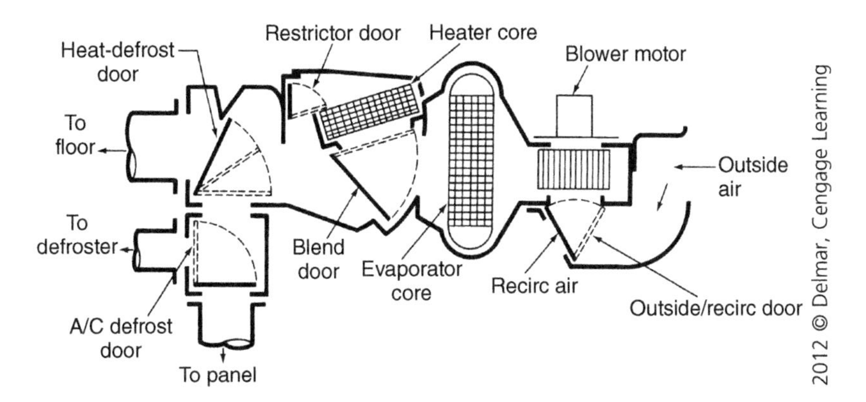

42. Referring to the figure above, any of these statements about airflow through the duct box are correct EXCEPT:

A. The blend door routes air through or around the evaporator core.
B. The outside/recirc door position affects the origin of the air flowing through the box.
C. The A/C defrost door routes the air through to the defroster or to the panel vents.
D. The blower motor forces air through the passages and is controlled by a switch.

43. A technician connects a scan tool to a late-model vehicle to troubleshoot a problem in the automatic temperature control (ATC) system. The scan tool gives a fault of ambient air temperature sensor "open circuit." Technician A says that the problem could be a sensor wire rubbing a sharp body component. Technician B says that the problem could be a broken sensor wire. Who is correct?

A. A only
B. B only
C. Both A and B
D. Neither A nor B

44. Technician A says that a dual pressure switch serves two functions in the A/C system. Technician B says that a dual pressure switch will open if the system loses the refrigerant charge. Who is correct?

A. A only
B. B only
C. Both A and B
D. Neither A nor B

45. What function does the A/C clutch front drive plate perform for the A/C system?

    A. Prevents the voltage spike from damaging other components
    B. Assists the clutch coil in creating magnetism
    C. Limits current flow to prevent the A/C fuse from overheating
    D. Connects to the driven pulley when the coil is energized

46. A vehicle is being diagnosed for engine coolant loss. The cooling system is pressurized at 15 psi (103 kPa) for 15 minutes. There is no visible sign of coolant leaks in the engine or passenger compartments, but the pressure on the tester gauge decreases to 5 psi (34 kPa). This problem could be caused by any of the following defects EXCEPT:

    A. A leaking heater core
    B. A leaking internal transmission cooler
    C. A leaking head gasket
    D. A cracked cylinder head

47. Technician A says that the service advisor needs to be certified by SAE in order to write repair orders for A/C repairs. Technician B says that the shop owner must be certified by a recognized body on how to properly handle refrigerants. Who is correct?

    A. A only
    B. B only
    C. Both A and B
    D. Neither A nor B

48. A refrigerant identifier is connected to an A/C system and gives the reading of 100 percent R134a and 0 percent R12. Technician A says that this system can be safely recovered into the R134a recovery machine. Technician B says that refrigerant identifiers should be used on every vehicle prior to connecting any A/C equipment. Who is correct?

    A. A only
    B. B only
    C. Both A and B
    D. Neither A nor B

49. The accumulator drier is being replaced on a late-model vehicle. Technician A says that the accumulator should be installed prior to the other A/C components. Technician B says that some refrigerant oil should be added to the accumulator prior to installation. Who is correct?

    A. A only
    B. B only
    C. Both A and B
    D. Neither A nor B

50. Technician A says that R134a systems use PAG oil to lubricate the compressor. Technician B says that R12 systems use mineral oil to lubricate the compressor. Who is correct?

    A. A only
    B. B only
    C. Both A and B
    D. Neither A nor B

PREPARATION EXAM 4

1. Any of these items are typically found in a wiring diagram EXCEPT:

   A. Splice locations
   B. Circuit number of a wire
   C. Connector numbers for the circuit in question
   D. Color of the wire

   Answer A is correct. Wiring diagrams do not typically describe the location of the splices that are used in the circuit.

   Answer B is incorrect. Wiring diagrams will usually show the circuit numbers of the wires involved in the circuit.

   Answer C is incorrect. Wiring diagrams usually show the connector numbers as well as pins identifications of the wires involved in the circuit.

   Answer D is incorrect. Wiring diagrams will usually show the colors of the wires involved in the circuit.

2. The air gap on a compressor clutch is found to be 0.003 inches. The specification for this compressor is 0.020 to 0.030 inches. Technician A says this could cause an intermittent scraping noise with the engine running and the compressor clutch engaged. Technician B says this could cause a slipping compressor clutch. Who is correct?

   A. A only
   B. B only
   C. Both A and B
   D. Neither A nor B

3. A customer states that the heater temperature is very inconsistent. It will warm up for a few minutes and then it will cool off for a few minutes. Technician A says that the coolant level should be checked for the correct level. Technician B says that a blown head gasket could cause this problem. Who is correct?

   A. A only
   B. B only
   C. Both A and B
   D. Neither A nor B

4. The air passages through an A/C condenser are severely restricted. Technician A says this may cause refrigerant discharge from the high-pressure relief valve. Technician B says this may cause the high-side pressure and low-side pressure to be lower than normal. Who is correct?

   A. A only
   B. B only
   C. Both A and B
   D. Neither A nor B

5. Technician A says when the engine is running with the thermostat closed the coolant is directed through a by-pass system to ensure coolant circulation. Technician B says some systems use the upper radiator hose as a thermostat by-pass. Who is correct?

   A. A only
   B. B only
   C. Both A and B
   D. Neither A nor B

6. Technician A says a loose ground wire could cause the blower fuse to burn up. Technician B says an inoperative blower could be caused by a defective compressor clutch diode. Who is correct?

   A. A only
   B. B only
   C. Both A and B
   D. Neither A nor B

7. What is the LEAST LIKELY cause of a failed electric mode actuator on an ATC-equipped vehicle?

   A. Binding blend door
   B. Stuck defrost door
   C. Binding heater door
   D. Stuck vent door

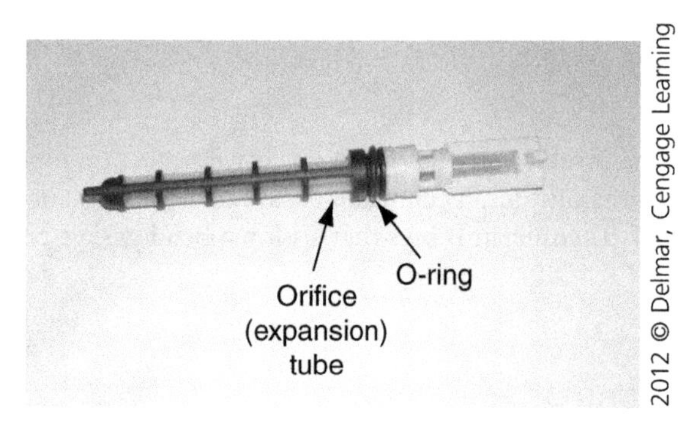

8. In the figure above, what function does the device perform for the refrigerant system?

   A. Meters refrigerant into the compressor
   B. Filters refrigerant into the condenser
   C. Meters refrigerant into the evaporator core
   D. Filters refrigerant into the compressor

9. Technician A says that a short to ground that is after the load will not cause a fuse to blow. Technician B says that an open circuit after the load will cause excessive voltage drop at the load of the circuit. Who is correct?

   A. A only
   B. B only
   C. Both A and B
   D. Neither A nor B

10. A vehicle is being diagnosed that had an alternative refrigerant added on the last service. Technician A says the system must be purged of all refrigerant to the atmosphere before it can be evaluated. Technician B says that a refrigerant identifier should be used to determine the type of refrigerant. Who is correct?

    A. A only
    B. B only
    C. Both A and B
    D. Neither A nor B

11. A blower motor operates only on high speed. The LEAST LIKELY cause of this condition would be?

    A. Blower motor ground is open
    B. Open wire near the blower resistor
    C. Blower switch
    D. Blower resistor

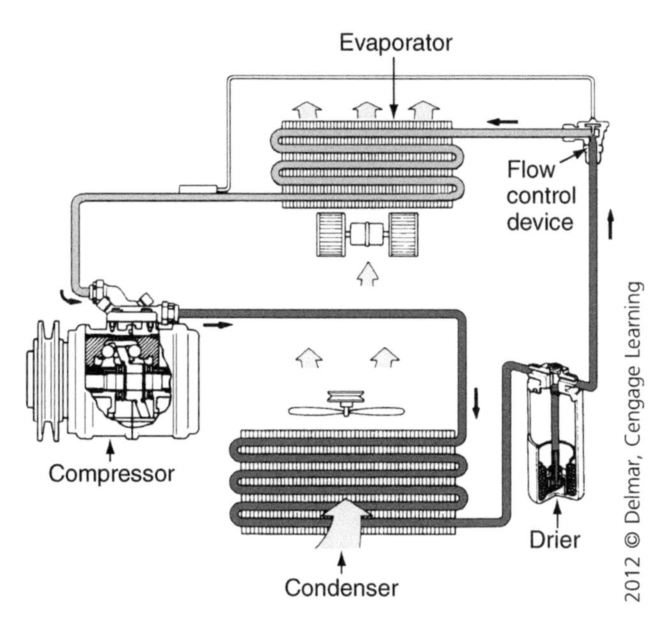

12. Referring to the figure above, what is the pressure and state of the refrigerant as it leaves the compressor?

    A. High-pressure liquid
    B. Low-pressure vapor
    C. High-pressure vapor
    D. Low-pressure liquid

13. Technician A says the lines entering and exiting the receiver/drier should be of the same temperature. Technician B says the lines before and after the condenser should not be of the same temperature. Who is correct?

    A. A only
    B. B only
    C. Both A and B
    D. Neither A nor B

14. Technician A says the A/C system can be charged through the low side with the system running. Technician B says inverting the refrigerant container causes low-pressure refrigerant vapor to be charged into the system. Who is correct?

    A. A only
    B. B only
    C. Both A and B
    D. Neither A nor B

15. Any of these components provide inputs to the automatic temperature control (ATC) computer EXCEPT:

    A. Sunlight sensor
    B. Engine speed sensor
    C. Cabin temperature sensor
    D. Ambient temperature sensor

16. A vehicle makes a hissing noise each time the A/C system and engine are turned off. Technician A says that the noise is caused by a coolant leak. Technician B says that the noise is caused by equalization of system pressures. Who is correct?

    A. A only
    B. B only
    C. Both A and B
    D. Neither A nor B

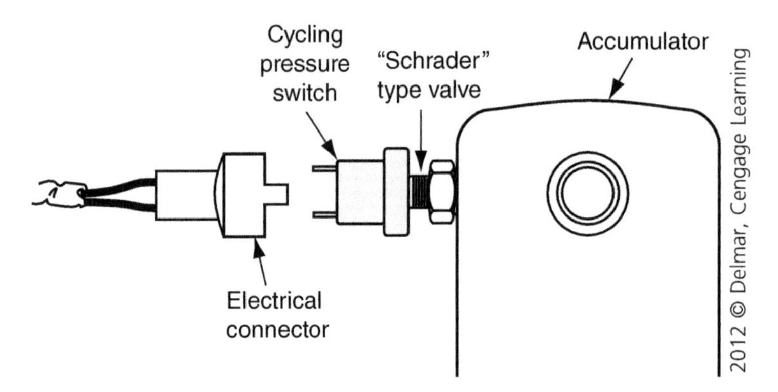

17. Referring to the figure above, Technician A says that the refrigerant will need to be recovered prior to removing the pressure cycling switch. Technician B says that the accumulator is used on TXV-type systems. Who is correct?

    A. A only
    B. B only
    C. Both A and B
    D. Neither A nor B

18. A vehicle is being diagnosed for higher-than-normal system pressures. Technician A checks the condenser for dirt buildup in the condenser's fins. Technician B tests the engine for indications of an overheating condition. Who is correct?

    A. A only
    B. B only
    C. Both A and B
    D. Neither A nor B

19. Any of these conditions must be present when charging an A/C system into the high side EXCEPT:

    A. The system must be off.
    B. The system should be under a vacuum.
    C. The refrigerant container should be inverted for liquid charging.
    D. The refrigerant container should be heated with a torch.

20. Technician A says that a loose air conditioning belt typically squeals as the compressor clutch is engaged. Technician B says that a slipping serpentine belt can be caused by a defective automatic tensioner. Who is correct?

    A. A only
    B. B only
    C. Both A and B
    D. Neither A nor B

21. Technician A says that some A/C systems control evaporator pressure by varying the displacement of the compressor. Technician B says some systems regulate the flow of refrigerant into the evaporator with a thermal expansion valve (TXV). Who is correct?

    A. A only
    B. B only
    C. Both A and B
    D. Neither A nor B

22. During a performance test of a TXV-style refrigerant system, the high-side pressure is excessively high and there is a frosted area on the condenser near the outlet. Technician A says the cause may be a refrigerant overcharge. Technician B says there may be a restriction in the condenser. Who is correct?

    A. A only
    B. B only
    C. Both A and B
    D. Neither A nor B

23. Any of these statements about maintaining certified A/C equipment are correct EXCEPT:

    A. Equipment must be UL approved.
    B. Equipment must have an approved storage container.
    C. Equipment must be capable of mixing R12 and R134a.
    D. Equipment must meet SAE standards.

24. Any of these A/C components can be flushed during A/C service EXCEPT:

 A. Muffler
 B. A/C suction hose
 C. A/C liquid line
 D. Evaporator core

25. A temperature drop test is being performed on a condenser. The inlet line was measured at 145°F and the outlet line temperature was measured at 120°F. Technician A says that the condenser is likely internally restricted. Technician B says that outlet temperature should be hotter than the inlet temperature. Who is correct?

 A. A only
 B. B only
 C. Both A and B
 D. Neither A nor B

26. Technician A says that a leaking A/C compressor shaft seal will not produce any visible sign of a leak. Technician B says that a compressor shaft seal can be replaced without recovering the refrigerant. Who is correct?

 A. A only
 B. B only
 C. Both A and B
 D. Neither A nor B

27. A vehicle is being diagnosed for a poor performing heater. The HVAC control is set to full heat but the temperature of the air is barely warm. Technician A says the heater core may be restricted. Technician B says the engine cooling system thermostat may be stuck open. Who is correct?

 A. A only
 B. B only
 C. Both A and B
 D. Neither A nor B

28. There is a noticeable noise coming from the engine compartment when the A/C is selected. Technician A says this could be caused by a loose compressor mount. Technician B says this could be caused by the discharge line touching a metal bracket. Who is correct?

 A. A only
 B. B only
 C. Both A and B
 D. Neither A nor B

29. Any of these cooling system tests are valuable to perform during a maintenance service EXCEPT:

 A. pH test for acidity
 B. Freeze protection
 C. Coolant voltage
 D. Water pump end-play

30. Technician A says a purpose of the A/C clutch diode is to prevent alternating current from the alternator entering the clutch coil. Technician B says a purpose of the A/C clutch diode is to prevent spikes of high voltage created by the clutch coil operation from damaging delicate electronic components. Who is correct?

    A. A only
    B. B only
    C. Both A and B
    D. Neither A nor B

31. Technician A says that some vehicles use the throttle position sensor input to disengage the A/C compressor during times of heavy load. Technician B says some vehicles have a power steering cutoff switch to disengage the A/C compressor when the power steering requires maximum effort. Who is correct?

    A. A only
    B. B only
    C. Both A and B
    D. Neither A nor B

32. Technician A says that all of the duct air typically passes through the evaporator core. Technician B says the incoming air must go through the evaporator even if heat is selected. Who is correct?

    A. A only
    B. B only
    C. Both A and B
    D. Neither A nor B

33. Technician A says that an open circuit will typically cause a fuse to blow. Technician B says that a corroded connection will typically cause a fuse to blow. Who is correct?

    A. A only
    B. B only
    C. Both A and B
    D. Neither A nor B

34. What can a refrigerant identifier determine about a refrigerant system?

    A. Air in the refrigerant system
    B. Excessive refrigerant charge
    C. Excessive oil in the refrigerant system
    D. Low refrigerant charge

35. A blower motor switch is being inspected and the technician finds burned contacts on the switch. What would be the most likely cause of this condition?

    A. Open blower resistor
    B. Tight blower motor bearing
    C. Loose blower motor ground
    D. High resistance across the blower relay

36. What is the most likely tool to be used when troubleshooting a data communication network?

    A. Continuity tester
    B. Digital multi-meter (DMM)
    C. 12 volt test light
    D. Short finder

37. Technician A says that electric mode actuators use a position sensor to provide feedback to the HVAC computer about door position. Technician B says that electric blend door actuators use a position sensor to provide the HVAC computer information about the blend door position. Who is correct?

    A. A only
    B. B only
    C. Both A and B
    D. Neither A nor B

38. A hissing sound is heard behind the dash panel. Technician A says that the vacuum switching valve could be leaking. Technician B says that a vacuum hose could be pinched. Who is correct?

    A. A only
    B. B only
    C. Both A and B
    D. Neither A nor B

39. Any of these methods of troubleshooting vacuum problems on an HVAC duct system are correct EXCEPT:

    A. Listening for hissing sounds with an ultrasonic leak detector
    B. Using a vacuum pump to test vacuum actuators
    C. Connecting the A/C evacuation pump to the system to monitor leaks
    D. Visually inspecting hoses and actuators

40. Technician A says that failing to change the cabin air filter according to the maintenance schedule could result in reduced airflow from the ducts. Technician B says that the cabin air filters should be serviced every 100,000 miles. Who is correct?

    A. A only
    B. B only
    C. Both A and B
    D. Neither A nor B

41. Technician A says that some automatic temperature control (ATC) systems have a self-diagnostic feature in the HVAC control head. Technician B says that some automatic temperature control (ATC) systems have a calibration feature available in the HVAC control head. Who is correct?

    A. A only
    B. B only
    C. Both A and B

42. Technician A says that some scan tools have an output test function that will energize the A/C compressor clutch. Technician B says that some scan tools have an output test that will energize the A/C pressure cycling switch. Who is correct?

    A. A only
    B. B only
    C. Both A and B
    D. Neither A nor B

43. What is the LEAST LIKELY step that would need to be followed when replacing the A/C control panel?

    A. Disconnect the wiring and cables.
    B. Remove the trim bezel.
    C. Recover the refrigerant.
    D. Disconnect the negative battery cable.

44. Technician A says that the scan tool receives data from the vehicle by communicating on the data bus network. Technician B says that if both of the data bus wires break, then the network will not communicate. Who is correct?

    A. A only
    B. B only
    C. Both A and B
    D. Neither A nor B

45. Technician A says that a refrigerant identifier will determine the purity of the refrigerant. Technician B says that the refrigerant identifier should be connected to a vehicle prior to connecting the recovery station. Who is correct?

    A. A only
    B. B only
    C. Both A and B
    D. Neither A nor B

46. Any of these examples are possible methods for a technician to retrieve trouble codes from a vehicle HVAC system EXCEPT:

    A. Connect a scan tool to the vehicle data link connector (DLC) to communicate with the HVAC system.
    B. Connect a scan tool directly to the HVAC control module to communicate with the HVAC system.
    C. Depress the buttons on the HVAC control head and watch the flashing indicator.
    D. Depress the buttons on the HVAC control head and read the code on the electronic display.

    Answer A is incorrect. Trouble codes can be retrieved by connecting a scan tool to the DLC.

    Answer B is correct. The scan tool is not typically connected directly to the HVAC control module.

    Answer C is incorrect. Some vehicles will display trouble codes after buttons are depressed on the HVAC control head. The codes are displayed by a flashing indicator on the control head.

    Answer D is incorrect. Some vehicles will display trouble codes after buttons are depressed on the HVAC control head. The codes are displayed by an electronic display on the control head.

47. A pressure check of a tank of refrigerant that had been stored 16 hours at 70°F (21.1°C) showed 98 psi (673.95 kPa). What is the most likely factor that would cause these readings?

    A. Excessive refrigerant oil
    B. Container is overfilled
    C. Excessive air
    D. Normal reading

48. Technician A says the high-pressure relief valve is a mechanical relief valve that exhausts refrigerant when the pressure exceeds 280 psi (1930.5 kPa). Technician B says that the high-pressure relief valve resets automatically after it releases pressure. Who is correct?

    A. A only
    B. B only
    C. Both A and B
    D. Neither A nor B

    Answer A is incorrect. High-pressure relief valves typically will not release pressure until a much higher point than 280 psi (1930.5 kPa). Each manufacturer will have a specification, but none would be this low due to pressures normally running this high on very hot and humid days.

    Answer B is correct. Only Technician B is correct. The high-pressure relief valve is a mechanical blow-off valve that automatically resets after it releases pressure.

    Answer C is incorrect. Only Technician B is correct.

    Answer D is incorrect. Technician B is correct.

49. All of the vacuum controls are inoperative with the engine running at idle speed. Technician A says the problem could be caused by a blocked vacuum supply hose. Technician B says the manifold vacuum fitting may be disconnected. Who is correct?

    A. A only
    B. B only
    C. Both A and B
    D. Neither A nor B

50. Which of these processes removes the refrigerant from the A/C system?

    A. Identification
    B. Leak test
    C. Recovery
    D. Evacuation

PREPARATION EXAM 5

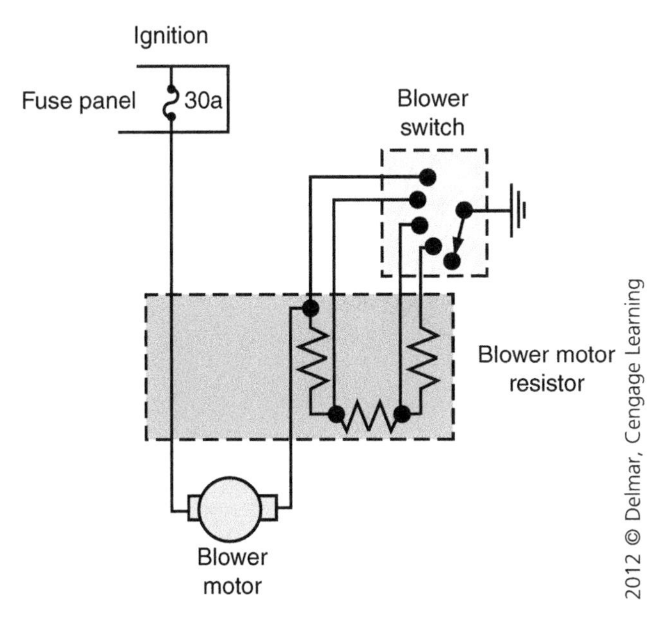

1. How many speeds does the blower motor in the figure above have?
   A. Three
   B. Four
   C. Five
   D. Six

2. Technician A says the lines before and after the receiver/drier should be of the same temperature. Technician B says the lines before and after the condenser should be of the same temperature. Who is correct?
   A. A only
   B. B only
   C. Both A and B
   D. Neither A nor B

3. An A/C system has a vacuum reservoir, a check valve, and a vacuum-operated mode door. While operating in the A/C mode and climbing a steep hill with the throttle nearly wide open, the air discharge switches from the panel to the defroster ducts. The A/C system operates normally under all other conditions. The most likely cause of this problem would be:
   A. A leaking panel door vacuum actuator
   B. A defective vacuum reservoir check valve
   C. A leaking blend-air door vacuum actuator
   D. A leaking intake manifold gasket

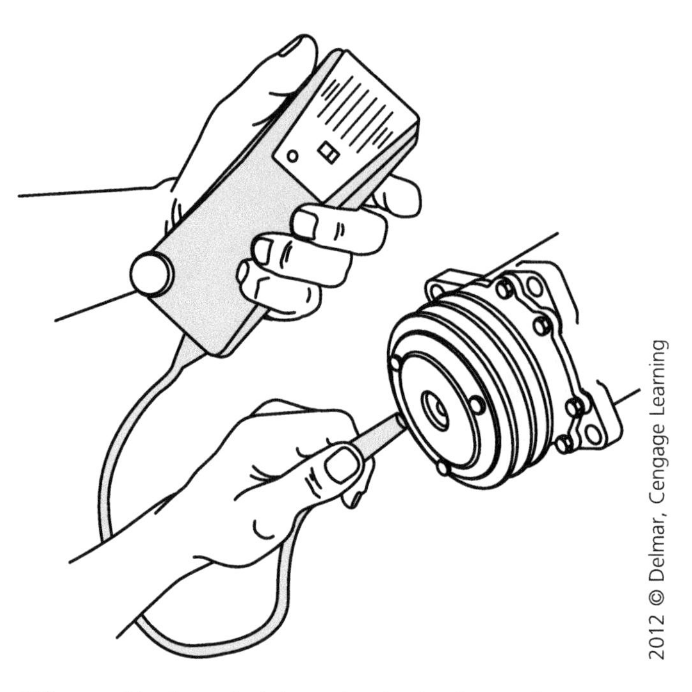

4. What tool is the technician using in the figure above?
   - A. Refrigerant identifier
   - B. Belt tension gauge
   - C. Refrigerant leak detector
   - D. Electronic manifold device

5. Any of these faults in the A/C system could cause an elevated high-side reading EXCEPT:
   - A. Refrigerant overcharge
   - B. Restricted airflow to the condenser
   - C. Poor airflow across the evaporator
   - D. A slipping fan clutch

6. A vehicle being serviced is identified to have an alternative refrigerant in the A/C system. Technician A says the system should be evacuated before service begins. Technician B says that the refrigerant should be vented to the atmosphere and then recharged with pure refrigerant. Who is correct?
   - A. A only
   - B. B only
   - C. Both A and B
   - D. Neither A nor B

7. Technician A says that some A/C systems prevent the evaporator from freezing up by varying the displacement of the compressor. Technician B says some systems regulate the flow of refrigerant into the evaporator with a variable orifice valve (VOV). Who is correct?
   - A. A only
   - B. B only
   - C. Both A and B
   - D. Neither A nor B

8. Technician A says a vacuum pump is used during the recovery process to remove the last trace of refrigerant from the system. Technician B says any oil not recovered from the air conditioning system reduces the system's efficiency. Who is correct?

   A. A only
   B. B only
   C. Both A and B
   D. Neither A nor B

9. An automatic temperature control (ATC) A/C system displays a diagnostic trouble code (DTC) fault for the temperature blend door actuator motor. Technician A says the first step in the repair procedure is to replace the temperature blend door actuator. Technician B says that a diagnostic flow chart usually needs to be followed in order to find the problem. Who is correct?

   A. A only
   B. B only
   C. Both A and B
   D. Neither A nor B

10. An A/C thermal limiter switch is being diagnosed. Technician A says a thermal switch can be connected in series with the compressor clutch. Technician B says a thermal switch is usually mounted on the compressor. Who is correct?

    A. A only
    B. B only
    C. Both A and B
    D. Neither A nor B

11. Technician A says a replacement compressor should have the same type of line connection points as the old compressor. Technician B says the mounting brackets and other fasteners on a replacement compressor should be identical to those on the old compressor. Who is correct?

    A. A only
    B. B only
    C. Both A and B
    D. Neither A nor B

12. Technician A says if an air conditioning system has too much refrigerant oil, the performance of the system will suffer. Technician B says poor system performance can be caused by an overcharge of refrigerant in the system. Who is correct?

    A. A only
    B. B only
    C. Both A and B
    D. Neither A nor B

13. Technician A says that restricted refrigerant passages in the evaporator may cause frosting of the evaporator outlet pipe. Technician B says restricted refrigerant passages in the evaporator may cause much higher-than-specified low-side pressures. Who is correct?

    A. A only
    B. B only
    C. Both A and B
    D. Neither A nor B

14. A performance test reveals the pressure on the low side of the refrigerant system is higher than specification and the pressure on the high side of the refrigerant system is lower than specification. Technician A says that the A/C condenser could be restricted internally. Technician B says that the intake reed valve in the compressor could be broken. Who is correct?

    A. A only
    B. B only
    C. Both A and B
    D. Neither A nor B

15. An A/C compressor is being diagnosed for a slipping front clutch drive plate. A voltage test at the compressor clutch coil shows 8.5 volts when the compressor is engaged. Technician A says that the A/C clutch relay could have high resistance on the load-side contacts. Technician B says that the A/C coil ground could be loose. Who is correct?

    A. A only
    B. B only
    C. Both A and B
    D. Neither A nor B

16. Any of these conditions should cause the electric cooling fan relay to energize the fan motor EXCEPT:

    A. Coolant temperature above 230°F
    B. Check engine light blinking with engine running
    C. An output test using a scan tool
    D. A/C refrigerant pressure above 400 psi

17. Technician A says that some vacuum leaks in the HVAC system can be diagnosed with an ultrasonic leak detector. Technician B says that some mechanical HVAC switching systems can be diagnosed by listening to the doors hit the limits while moving the control head back and forth. Who is correct?

    A. A only
    B. B only
    C. Both A and B
    D. Neither A nor B

18. The airflow at the condenser is severely restricted with dirt and mud. Technician A says that the duct temperature will be higher than normal on a hot day. Technician B says that system pressures will be elevated. Who is correct?

    A. A only
    B. B only
    C. Both A and B
    D. Neither A nor B

19. Which device is used as the input to cause the A/C compressor clutch to disengage during a parallel parking maneuver?

    A. Coolant temperature sensor
    B. Throttle position sensor
    C. Oxygen sensor
    D. Power steering switch

20. While testing the A/C system, the low-side pressure is found to be higher than normal. Cooling the expansion valve remote bulb lowers the low-side pressure. Technician A says the expansion valve could be defective. Technician B says the remote bulb may be improperly secured to the evaporator outlet. Who is correct?

    A. A only
    B. B only
    C. Both A and B
    D. Neither A nor B

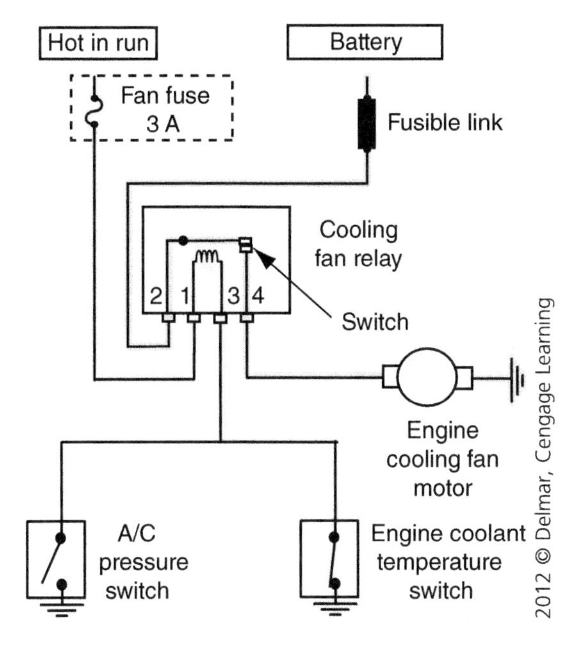

21. Referring to the figure above, the engine cooling fan does not operate when sitting in traffic for long periods of time but it will come on when the A/C is turned on. Technician A says that the fusible link could be open. Technician B says that the low-speed fan relay coil could be shorted. Who is correct?

    A. A only
    B. B only
    C. Both A and B
    D. Neither A nor B

22. The vacuum valve in the radiator cap is stuck closed. The result of this problem could be:

    A. Collapsed upper radiator hose after the engine was shut off
    B. Excessive cooling system pressure at normal engine temperature
    C. Engine overheating when operating under a heavy load
    D. Engine overheating during extended idle periods

23. An automatic temperature control (ATC) processor needs to be replaced. Technician A says that the new processor may need to be reprogrammed with a scan tool. Technician B says that the HVAC control head will have to be replaced at the same time. Who is correct?

    A. A only
    B. B only
    C. Both A and B
    D. Neither A nor B

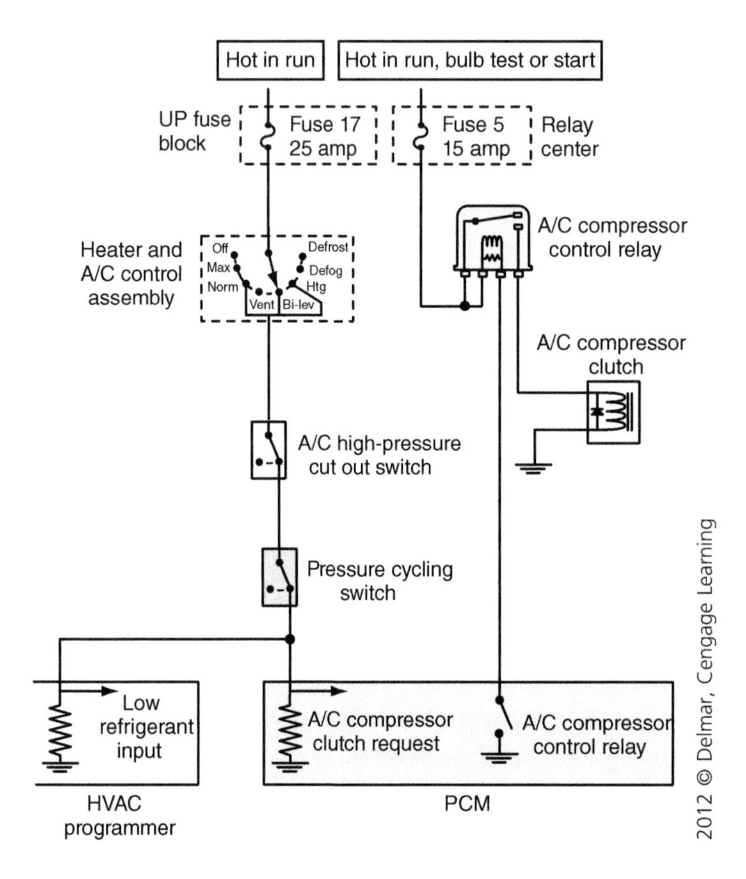

24. Referring to the figure above, the A/C compressor clutch will not engage when the A/C is selected at the HVAC control, but the compressor clutch will engage when commanding it on with a scan tool function output test. Technician A says that the HVAC control head could have an internal fault. Technician B says that the 15 amp fuse in the relay center could be open. Who is correct?

    A. A only
    B. B only
    C. Both A and B
    D. Neither A nor B

25. The heater water valve is being checked. The hose between the control valve and heater core is hot, and the outlet hose from the heater core is much cooler. Technician A says the coolant control valve may be defective and should be cleaned or replaced. Technician B says the heater core may be clogged and should be cleaned or replaced. Who is correct?

    A. A only
    B. B only
    C. Both A and B
    D. Neither A nor B

26. Any of these statements about computer-controlled A/C system duct door actuator motors are true EXCEPT:

    A. Some actuator motors are calibrated automatically using the self-diagnostic mode of the control head.
    B. AC system and component problems sometimes produce diagnostic trouble codes.
    C. The control rods must be calibrated manually on some systems.
    D. The actuator motor may require calibration after motor replacement.

27. Technician A says if the refrigerant is contaminated with moisture, chemical changes in the refrigerant can cause corrosion. Technician B says if refrigerant oil is allowed to mix with the refrigerant the system will not build up the necessary pressures to work efficiently. Who is correct?

    A. A only
    B. B only
    C. Both A and B
    D. Neither A nor B

28. Technician A says that a blower resistor is used to lower voltage to the blower motor to create different speeds. Technician B says that the resistance provided by the blower resistor is highest at the lowest blower speed. Who is correct?

    A. A only
    B. B only
    C. Both A and B
    D. Neither A nor B

29. The compressor clutch diode is being diagnosed. Technician A says a purpose of the diode is to control current flow through the A/C clutch coil when the coil is energized. Technician B says a purpose of the diode is to prevent spikes of high voltage created by the clutch coil operation from damaging delicate electronic components. Who is correct?

    A. A only
    B. B only
    C. Both A and B
    D. Neither A nor B

30. A car is equipped with an A/C compressor driven by a serpentine belt. As the compressor cycles, the belt squeals briefly. Technician A says the belt tensioner may be weak. Technician B says that the system may be undercharged with refrigerant. Who is correct?

    A. A only
    B. B only
    C. Both A and B
    D. Neither A nor B

31. Any of these statements concerning the refrigeration system are true EXCEPT:

    A. Refrigerant leaves the compressor as a high-pressure, high-temperature vapor.
    B. Impurities are removed by the receiver/drier.
    C. The expansion valve controls the flow of refrigerant into the evaporator core.
    D. Warm air passing through the evaporator causes heat to be absorbed into the refrigerant.

32. A scan tool can be used for any of these actions EXCEPT:

    A. Retrieving diagnostic trouble codes (DTCs)
    B. Repairing a poor terminal connection at a sensor
    C. Performing output tests on actuators that are controlled by a control module
    D. Viewing live data from sensors that are inputs to a control module control modules.

33. The liquid line has frost on it right at the point that exits the receiver/drier. Technician A says that the receiver/drier could be restricted. Technician B says that the compressor could have faulty reed valves. Who is correct?

    A. A only
    B. B only
    C. Both A and B
    D. Neither A nor B

34. The blower motor does not operate on high speed. All other speeds are operational. Technician A says the blower ground wire may be loose or broken. Technician B says the high-speed relay may be defective. Who is correct?

    A. A only
    B. B only
    C. Both A and B
    D. Neither A nor B

35. An ambient temperature sensor is being diagnosed. Technician A says this device senses outside air temperature by varying the resistance as the outside light level changes. Technician B says that this device is typically a variable negative temperature coefficient (NTC) thermistor. Who is correct?

    A. A only
    B. B only
    C. Both A and B
    D. Neither A nor B

36. Which of these test tools would be the most likely option to use when testing the resistance of a cabin temperature sensor?
    A. Digital multi-meter
    B. Scan tool
    C. Digital storage oscilloscope (DSO)
    D. Jumper wire

37. The low-side pressure is lower than normal. Which of these is the LEAST LIKELY cause?
    A. A faulty metering device
    B. Poor airflow across the evaporator
    C. A restriction in the low side of the system
    D. System overcharged with refrigerant

38. Quick-connect refrigerant lines are being serviced. For which process are special tools required?
    A. Disconnecting spring lock couplings
    B. Connecting spring lock couplings
    C. Installing the o-rings
    D. Inspecting the couplings

39. An automatic temperature control (ATC) problem is being diagnosed. The scan tool gives a fault of "data bus open." Technician A says that the problem could be a data wire rubbing a sharp body component. Technician B says that the problem could be a broken data wire. Who is correct?
    A. A only
    B. B only
    C. Both A and B
    D. Neither A nor B

40. The blower motor fuse blows after the blower is operated on high speed for several minutes. The fuse is replaced and blows again after the same amount of time with the blower on high. Technician A says that the problem is likely a direct short to ground before the load in blower circuit. Technician B says that the blower motor could have a tight bearing. Who is correct?
    A. A only
    B. B only
    C. Both A and B
    D. Neither A nor B

41. A vehicle with automatic temperature control (ATC) is malfunctioning. A scan tool is connected and a code is retrieved from the system that indicates a fault in the sun load sensor. Technician A says that the sensor should be tested with a test light. Technician B says the sensor can be checked with a digital ohmmeter to see if it varies resistance when a flashlight is shined onto it. Who is correct?
    A. A only
    B. B only
    C. Both A and B
    D. Neither A nor B

42. Technician A says that the computer senses the position of the mode door by monitoring the signal voltage of the position sensor. Technician B says that the computer provides the ground for the position sensor. Who is correct?

A. A only
B. B only
C. Both A and B
D. Neither A nor B

43. Technician A says that the o-rings should be reused on most A/C repairs. Technician B says that the accumulation of oily residue around an A/C line connection is an indication of a leak in the system. Who is correct?

A. A only
B. B only
C. Both A and B
D. Neither A nor B

44. While testing a compressor clutch, a fused jumper wire is used to by-pass the load side of the A/C clutch relay and the compressor clutch engages. The most likely problem in this system would be:

A. An open compressor clutch coil
B. An open pressure cycling switch
C. An open compressor clutch coil ground circuit
D. An open relay power feed circuit

45. Technician A says it is common to mix R-12 and R-134a when servicing late-model A/C systems. Technician B says that some A/C service machines can recover, evacuate, and recharge A/C systems. Who is correct?

A. A only
B. B only
C. Both A and B
D. Neither A nor B

46. Technician A says that the refrigerant should be recovered before connecting a refrigerant identifier. Technician B says that a refrigerant identifier can test for flammable refrigerant substances. Who is correct?

A. A only
B. B only
C. Both A and B
D. Neither A nor B

47. Technician A says that some data communication network wires can be repaired by carefully soldering the wires. Technician B says that some data network wiring failures require the use of a digital multi-meter to diagnose. Who is correct?

A. A only
B. B only
C. Both A and B
D. Neither A nor B

48. A stuck open engine cooling system thermostat could cause any of these symptoms EXCEPT:

    A. Poor heater operation
    B. Below-normal engine operating temperature
    C. Increased fuel consumption
    D. Coolant loss

49. A vehicle with dual-zone climate control is being diagnosed. Technician A says that these systems use two separate duct boxes to deliver the varied airflow to the driver and passenger. Technician B says that these systems use two separate cabin air temperature sensors to provide feedback from each side of the car. Who is correct?

    A. A only
    B. B only
    C. Both A and B
    D. Neither A nor B

50. Technician A says that the compressor clutch coil can be tested with a digital ohmmeter. Technician B says that some compressor clutch coils have a diode wired in parallel with the coil. Who is correct?

    A. A only
    B. B only
    C. Both A and B
    D. Neither A nor B

PREPARATION EXAM 6

1. Technician A says that the refrigerant enters the compressor as a low-pressure gas. Technician B says that the refrigerant leaves the compressor as a high-pressure liquid. Who is correct?

    A. A only
    B. B only
    C. Both A and B
    D. Neither A nor B

2. Which of these statements is LEAST LIKELY to cause heating system problems?

    A. High concentration of antifreeze in the coolant
    B. Sticking or disconnected blend door
    C. Heater control valves stuck open
    D. Air pockets in the heater core

3. Any of these characteristics of a normal operating A/C system are correct EXCEPT:
   - A. The discharge line should be warm or hot when the A/C system is engaged.
   - B. The liquid line should be warm or hot when the A/C system is engaged.
   - C. The HVAC drain tube should produce water after several minutes of operation in warm conditions.
   - D. The components on the low side should be frosty when the A/C system is engaged.

4. The A/C compressor clutch relay is being diagnosed. Technician A says that the relay coil can be tested for an open with a digital ohmmeter. Technician B says that some A/C compressor clutch relays can be activated with an output test using a scan tool. Who is correct?
   - A. A only
   - B. B only
   - C. Both A and B
   - D. Neither A nor B

5. Any of these statements about the tools needed to service the refrigerant system are true EXCEPT:
   - A. Separate gauges and other refrigerant-handling equipment should be used for different types of refrigerant.
   - B. Manifold gauge sets for R-134a can be identified by unique service fitting connections.
   - C. R-134a pressure gauges are much stronger than R-12 gauges.
   - D. Proper identification of service equipment and hoses is important to prevent cross contamination of refrigerant.

6. Technician A says that the dual pressure switch prevents the compressor from operating if the system has lost all of the refrigerant. Technician B says that the dual pressure switch can cause the compressor to turn off if the high-side pressure exceeds specifications. Who is correct?
   - A. A only
   - B. B only
   - C. Both A and B
   - D. Neither A nor B

7. An A/C system performance test is performed on a late-model car. Technician A inserts a thermometer into the air duct at the center of the dash and monitors discharge temperature. Technician B connects a manifold pressure gauge set to the service fittings to monitor pressure. Who is correct?
   - A. A only
   - B. B only
   - C. Both A and B
   - D. Neither A nor B

8. The suction line is covered with a thick frost. Technician A says that this might indicate that the expansion valve is flooding the evaporator. Technician B says that the evaporator core may have a blockage. Who is correct?

   A. A only
   B. B only
   C. Both A and B
   D. Neither A nor B

9. A refrigerant identifier will detect any of these contaminants in an A/C system EXCEPT:

   A. Air
   B. Mixed refrigerants
   C. Sealer additive
   D. Flammable substance

10. Technician A says that individual parts of an electronic HVAC control panel can be tested and replaced. Technician B says that electronic HVAC control panels should be tested with a test light for correct operation. Who is correct?

    A. A only
    B. B only
    C. Both A and B
    D. Neither A nor B

11. Technician A says a container of PAG refrigerant oil must be kept closed when not in use to prevent the oil from absorbing moisture. Technician B says that any time an A/C system has a component replaced, oil must be added. Who is correct?

    A. A only
    B. B only
    C. Both A and B
    D. Neither A nor B

12. Any of these statements are correct about cabin air filter replacement EXCEPT:

    A. A clogged filter will not produce any noticeable problem in the HVAC system.
    B. The filter reduces the amount of dust and pollen from the passenger compartment.
    C. Driving conditions and terrain will alter the service intervals for the filter.
    D. A visual inspection is required to evaluate the need for replacement of the filter.

13. An evaporator core replacement is being performed on a late-model vehicle. Technician A disconnects the negative battery cable before working around the airbag components. Technician B stores the airbag components "face up" in a safe area while they are removed from the vehicle. Who is correct?

    A. A only
    B. B only
    C. Both A and B
    D. Neither A nor B

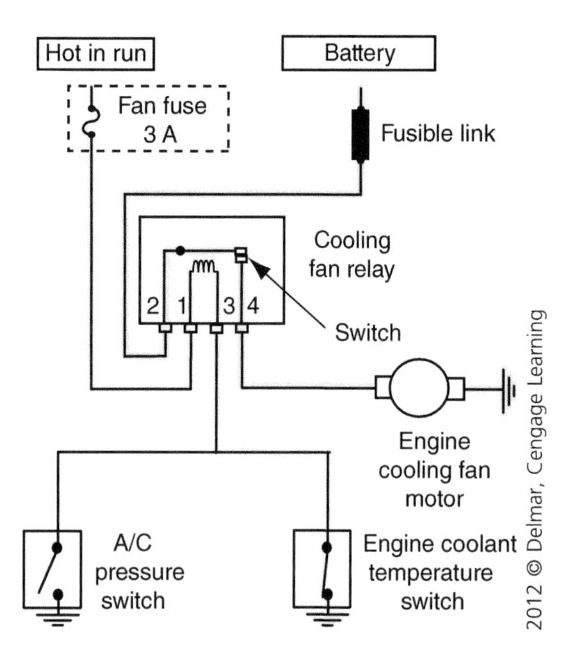

14. Referring to the figure above, the engine cooling fan runs constantly with the key in the run position. Technician A says that the A/C pressure switch could stuck be open. Technician B says that the low-speed fan relay load-side contacts could be shorted. Who is correct?

    A. A only
    B. B only
    C. Both A and B
    D. Neither A nor B

15. The serpentine drive belt is being inspected during an A/C repair. The inside of the belt is cracked and has chunks missing every 2 inches. Technician A says that the belt should be replaced. Technician B says that the belt tensioner should be replaced every time the belt is replaced. Who is correct?

    A. A only
    B. B only
    C. Both A and B
    D. Neither A nor B

16. Technician A says that the ATC control head can be used on some vehicles to retrieve diagnostic trouble codes from the HVAC computer. Technician B says that an electronic scan tool can be used to retrieve ATC diagnostic trouble codes from a vehicle. Who is correct?

    A. A only
    B. B only
    C. Both A and B
    D. Neither A nor B

17. Any of these statements about refrigerant oil are correct EXCEPT:
    A. PAG oil is the recommended lubricant for R-134a A/C systems.
    B. The system oil level can be checked with a dipstick on late-model A/C systems.
    C. Mineral oil is the recommended lubricant for R-12 A/C systems.
    D. PAG oil is a synthetic substance.

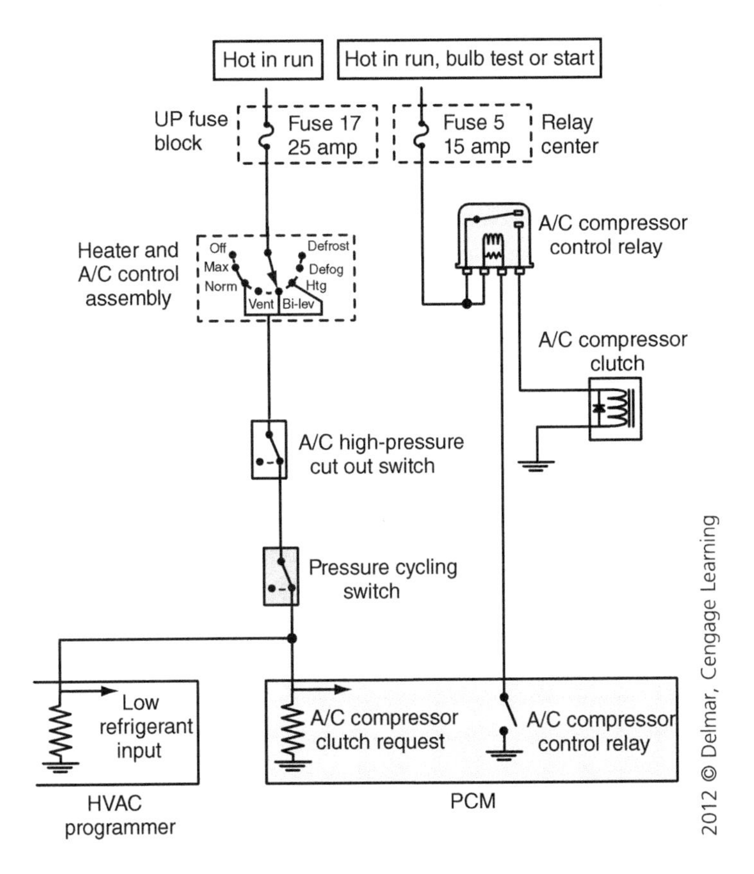

18. Referring to the figure above, the A/C compressor clutch will not engage when the A/C is selected on the HVAC control head. The technician attempts to engage the compressor clutch using a scan tool output test but it still does not work. Technician A says that the high-pressure cutout switch could be faulty. Technician B says that the A/C clutch diode could be open. Who is correct?
    A. A only
    B. B only
    C. Both A and B
    D. Neither A nor B

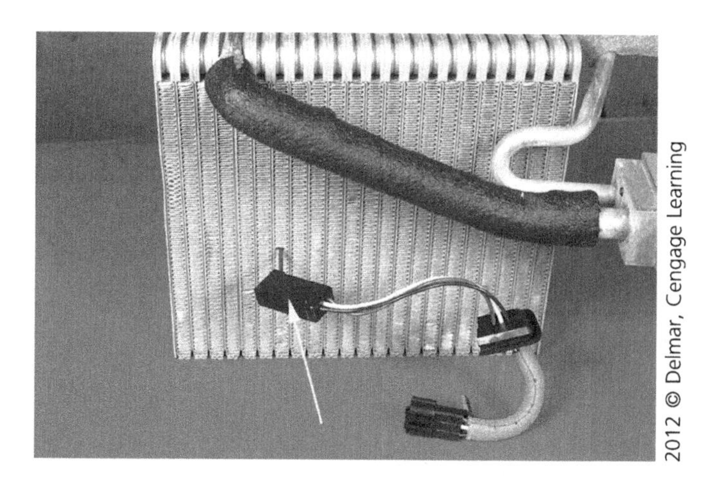

19. The component with the wire in the figure above could be any of these components EXCEPT:

    A. Evaporator temperature sensor
    B. Thermal expansion valve
    C. Thermostatic switch
    D. Fin temperature probe

20. The A/C suction/discharge hose is being replaced on a late-model vehicle. Technician A says that o-rings used on the connections should be installed dry and clean. Technician B says that the metal sealing washers should be installed dry and clean. Who is correct?

    A. A only
    B. B only
    C. Both A and B
    D. Neither A nor B

21. Which of these problems would be the most likely result of a condenser that is covered with debris?

    A. The discharge line would be warm to the touch.
    B. The suction line would have condensation dripping off of it.
    C. The accumulator would have condensation dripping off of it.
    D. The high-side pressure would be elevated.

22. An orifice tube is found to have heavy debris lodged on the screen. Technician A says that the screen can be cleaned and the orifice tube can be reused. Technician B says that the A/C system will need to be flushed and inspected to assure that all of the debris is removed from the system. Who is correct?

    A. A only
    B. B only
    C. Both A and B
    D. Neither A nor B

23. A receiver/drier is being diagnosed. Technician A says the receiver/drier should be changed if the outlet is colder than the inlet. Technician B says the receiver/drier should be changed if the refrigerant system has been exposed to the atmosphere for a long period of time. Who is correct?

A. A only
B. B only
C. Both A and B
D. Neither A nor B

24. The evaporator housing water drain is clogged. Technician A says that this may cause water to leak onto the front floorboard of the vehicle. Technician B says that this may cause water droplets to be present on the A/C duct vents when the system is operated. Who is correct?

A. A only
B. B only
C. Both A and B
D. Neither A nor B

25. Technician A says that the coolant freeze protection can be checked with a voltmeter. Technician B says that coolant freeze protection can be checked with a refractometer. Who is correct?

A. A only
B. B only
C. Both A and B
D. Neither A nor B

26. The temperature is measured at the heater ducts with the HVAC control set at full heat after the vehicle has been run for 20 minutes. The duct temperature is found to be only 90°F. Technician A says that the engine thermostat should be checked for correct operation. Technician B says that the heater core may be restricted. Who is correct?

A. A only
B. B only
C. Both A and B
D. Neither A nor B

27. A vehicle has an A/C compressor that only runs for a few seconds at a time and then shuts off. Technician A says that the refrigerant may have too much refrigerant oil. Technician B says the system may be low on refrigerant. Who is correct?

A. A only
B. B only
C. Both A and B
D. Neither A nor B

28. A blown HVAC fuse can be caused by any of these EXCEPT:

A. A short to ground in the temperature blend actuator circuit
B. A damaged cabin air temperature sensor
C. A damaged wiring harness
D. A short to ground in blower motor circuit

29. A vehicle comes to a repair shop with an A/C complaint and the A/C system is found to be empty. Technician A says that nitrogen could be used to leak test the empty A/C system. Technician B says that an electronic leak detector could be used to leak test the empty A/C system. Who is correct?

    A. A only
    B. B only
    C. Both A and B
    D. Neither A nor B

30. Technician A says that the dual pressure switch prevents the compressor from operating if the system has lost all of the refrigerant. Technician B says that the dual pressure switch can cause the compressor to turn off if the high-side pressure exceeds specifications. Who is correct?

    A. A only
    B. B only
    C. Both A and B
    D. Neither A nor B

31. The engine temperature sensor is being diagnosed. Technician A says that the A/C compressor may be deactivated if the engine temperature rises too high. Technician B says that the engine temperature sensor may cause the check engine light to illuminate if it develops a problem. Who is correct?

    A. A only
    B. B only
    C. Both A and B
    D. Neither A nor B

32. A vehicle that uses electric HVAC duct actuators is being diagnosed for an HVAC air handling problem. The air from the ducts does not come out at the correct location. Technician A says that a digital voltmeter could be used to measure the position sensor voltage on the mode actuators. Technician B says that the mode doors may need to be calibrated with a scan tool. Who is correct?

    A. A only
    B. B only
    C. Both A and B
    D. Neither A nor B

33. What is the LEAST LIKELY cause of an inoperative condenser fan motor?

    A. A blown fan maxi-fuse
    B. A faulty engine thermostat
    C. A faulty engine control module
    D. A faulty A/C pressure sensor

34. The blower motor operates slower than normal at all speed settings. A voltage test is performed at the blower connector with the switch in the high-speed switch position and 12.8 volts is measured. Technician A says that the cause could be a bad blower motor. Technician B says the problem could be a high resistance at the blower relay. Who is correct?

    A. A only
    B. B only
    C. Both A and B
    D. Neither A nor B

35. An A/C system needs to be recharged after a condenser replacement. Technician A says that the A/C system can be charged accurately by watching the gauge pressures. Technician B says that the A/C system can be charged accurately by watching the duct temperatures. Who is correct?

    A. A only
    B. B only
    C. Both A and B
    D. Neither A nor B

36. Technician A says that a dual-zone climate control system uses two blend doors. Technician B says that a dual-zone climate control system uses two fresh air actuators. Who is correct?

    A. A only
    B. B only
    C. Both A and B
    D. Neither A nor B

37. A vacuum-controlled HVAC system is being diagnosed. The air only comes out of the defrost duct, but the temperature and blower speeds operate normally. Technician A says that a kinked vacuum hose may be the cause. Technician B says that a broken vacuum hose may be the cause. Who is correct?

    A. A only
    B. B only
    C. Both A and B
    D. Neither A nor B

38. Any of these are common methods of moving the HVAC duct doors EXCEPT:

    A. Electric actuators
    B. Vacuum actuators
    C. Cables
    D. Air pressure

39. An HVAC duct door is being replaced on a late-model vehicle. Technician A says that the negative battery cable should be disconnected prior to beginning this repair. Technician B says that this repair can typically be made without opening up the duct box. Who is correct?

    A. A only
    B. B only
    C. Both A and B
    D. Neither A nor B

40. A vehicle with electronic dual-zone climate control has a problem of the driver's side only blowing cool air no matter what the driver setting is adjusted to. Technician A says that the driver's mode door actuator could be defective. Technician B says that the driver's blend door actuator could be defective. Who is correct?

A. A only
B. B only
C. Both A and B
D. Neither A nor B

41. The compressor clutch plate air gap is being inspected on a compressor. Technician A says that a feeler gauge can be used to make this measurement. Technician B says that the compressor clutch plate will slip when the compressor engages if this measurement is too wide. Who is correct?

A. A only
B. B only
C. Both A and B
D. Neither A nor B

42. A vacuum actuator is being tested by applying 18 in. Hg (40.5 kPa absolute) of vacuum to the vacuum actuator. Technician A says the vacuum gauge reading should remain steady for at least 1 minute. Technician B says if the gauge reading drops slowly, the actuator is leaking. Who is correct?

A. A only
B. B only
C. Both A and B
D. Neither A nor B

43. What function does the A/C clutch coil perform for the A/C system?

A. Prevents the voltage spike from damaging other components
B. Creates magnetism to attract the clutch drive plate
C. Limits current flow to prevent the A/C fuse from overheating
D. Protects the A/C system by opening and closing

44. A cable-controlled heater control valve is being diagnosed. The control knob is very difficult to move and the heater valve does move slowly. Technician A says the cable housing clamp may be loose at the control head (panel) end. Technician B says the cable may be rusted inside the cable housing. Who is correct?

A. A only
B. B only
C. Both A and B
D. Neither A nor B

45. Technician A says that some computer-controlled A/C system actuator motors are automatically calibrated in the self-diagnostic mode. Technician B says that some computer-controlled A/C system actuator motors require a scan tool for calibration after replacement. Who is correct?

A. A only
B. B only
C. Both A and B
D. Neither A nor B

46. Which of these is most likely to be performed during an A/C system recovery procedure?

A. Refrigerant is removed from the vehicle
B. Refrigerant is added to the vehicle
C. Most of the A/C system oil is removed the vehicle
D. Refrigerant is tested for contaminants

47. A mode actuator is being diagnosed on a late-model vehicle. Technician A says that a test light could be used to measure the voltage supplied to the actuator. Technician B says that a digital storage oscilloscope (DSO) could be used to detect an intermittent electrical fault at the actuator. Who is correct?

A. A only
B. B only
C. Both A and B
D. Neither A nor B

48. The refrigerant is being recovered from an A/C system. Five minutes after the recovery process is complete, the low-side pressure remains in a vacuum. This condition indicates:

A. There is still some refrigerant in the system.
B. There is excessive oil in the refrigerant system.
C. The refrigerant system is completely recovered.
D. There is excessive moisture in the refrigerant system.

49. A refrigerant identifier is connected to an A/C system and gives the reading of 99 percent R134a and 1 percent air. Technician A says that this system can be safely recovered into the R134a recovery machine. Technician B says that this system may have had previous repairs due to the small percentage of air. Who is correct?

A. A only
B. B only
C. Both A and B
D. Neither A nor B

50. Technician A says that the duct temperature will vary depending on the ambient temperature. Technician B says that the system pressures should vary with changes in ambient temperature. Who is correct?

A. A only
B. B only
C. Both A and B
D. Neither A nor B

SECTION

6

Answer Keys and Explanations

INTRODUCTION

Included in this section are the answer keys for each preparation exam, followed by individual, detailed answer explanations and a reference identifying the designated task area being assessed by each specific question. This additional reference information may prove useful if you need to refer back to the task list located in Section 4 of this book for additional support.

PREPARATION EXAM 1—ANSWER KEY

1. C
2. D
3. A
4. D
5. B
6. C
7. C
8. D
9. C
10. B
11. B
12. C
13. C
14. D
15. A
16. C
17. C
18. A
19. C
20. C
21. B
22. B
23. A
24. B
25. D
26. A
27. C
28. A
29. D
30. C
31. B
32. A
33. C
34. A
35. C
36. B
37. A
38. B
39. C
40. B
41. B
42. C
43. B
44. A
45. A
46. A
47. B
48. C
49. A
50. B

PREPARATION EXAM 1—EXPLANATIONS

1. An airbag system needs to be disarmed during an evaporator core replacement. Technician A says that it is necessary to remove power from the system prior to component removal. Technician B says that the inflatable devices should be laid "face up" in a secure area after being removed from a vehicle. Who is correct?

TASK A.11

   A. A only
   B. B only
   C. Both A and B
   D. Neither A nor B

   Answer A is incorrect. Technician B is also correct.

   Answer B is incorrect. Technician A is also correct.

   Answer C is correct. Both Technicians are correct. It is advisable to remove the power supply prior to airbag component removal. The best way to do this is to remove the negative battery cable. Live inflatable devices should be stored "face up" in a secure area to prevent the device from being accidentally deployed and then projected up in the air.

   Answer D is incorrect. Both Technicians are correct.

2. An orifice tube performs any of these jobs in the A/C system EXCEPT:

TASK B.2.6

   A. Meters refrigerant into the evaporator core
   B. Filters foreign particles
   C. Separates the high side from the low side
   D. Opens and closes to regulate pressure on the low side

   Answer A is incorrect. The orifice tube meters refrigerant into the evaporator core.

   Answer B is incorrect. The screen on the orifice tube helps to filter foreign particles from passing.

   Answer C is incorrect. The orifice tube is one of the separation points of the high and low side.

   Answer D is correct. The orifice does not open and close. It is a fixed restriction that meters refrigerant into the evaporator core.

3. A technician connects a scan tool to a late-model vehicle to troubleshoot a problem in the ATC system. The scan tool gives a fault of "data bus short to ground." Technician A says that the problem could be a data wire rubbing against a sharp body component. Technician B says that the problem could be a broken data wire. Who is correct?

TASK C.3.11

   A. A only
   B. B only
   C. Both A and B
   D. Neither A nor B

   Answer A is correct. Only Technician A is correct. A data wire that is rubbing against a sharp body component could cause a "short to ground" fault.

   Answer B is incorrect. A broken data wire would likely give an "open data wire" fault.

   Answer C is incorrect. Only Technician A is correct.

   Answer D is incorrect. Technician A is correct.

TASK A.4

4. Any of these methods of R134a A/C system leak detection are acceptable EXCEPT:
   A. Dye and black light
   B. Electronic leak detector
   C. Nitrogen and soapy water
   D. Propane burner

Answer A is incorrect. Injecting refrigerant dye and using a black light to inspect where the dye leaks out is a common method of leak detection.

Answer B is incorrect. Electronic leak detection is a very good method for A/C leak detection. The only disadvantage is that a technician is not able to reach the whole system with the probe.

Answer C is incorrect. Injecting compressed nitrogen and then using soapy water is an acceptable method to find A/C leaks.

Answer D is correct. The propane burner method is used as a leak detection method on R12 systems. It is not advisable to use this method on R134a systems.

TASK C.1.7

5. Technician A says that an electronic blend door motor uses an analog signal to open and close. Technician B says that an electronic mode door motor uses a feedback device to indicate the position of the door. Who is right?
   A. A only
   B. B only
   C. Both A and B
   D. Neither A nor B

Answer A is incorrect. Electronic blend door motors do not use an analog signal to operate. These motors typically receive a digital signal from an HVAC control device.

Answer B is correct. Only Technician B is correct. Electronic blend door motors use some sort of feedback device. A potentiometer is typically used to signal the HVAC control device the correct door position.

Answer C is incorrect. Only Technician B is correct.

Answer D is incorrect. Technician B is correct.

TASK B.1.3

6. What is the most likely cause for a serpentine drive belt to slip under heavy loads?
   A. Surface cracks on the inside of the belt
   B. Belt is stretched one-quarter inch
   C. Seized tensioner
   D. Glazed belt

Answer A is incorrect. Surface cracks on the inside of a serpentine belt are a sign of age, but they will not cause the belt to slip.

Answer B is incorrect. Belt stretch of one-quarter inch is not enough to cause a serpentine belt to slip. The spring-loaded tensioner will allow for some belt stretch.

Answer C is correct. A seized belt tensioner does not continue to adjust for belt wear and stretch.

Answer D is incorrect. A glazed belt could cause belt noise at various times but it does not typically cause the belt to slip.

7. Any of these practices should be followed when interacting with airbag system components EXCEPT:

TASK A.11

   A. Unhook the negative battery cable prior to beginning work.
   B. Store inflatable components "face up" in a secure area.
   C. Disconnect the airbag indicator prior to beginning work.
   D. Carefully carry inflatable components holding them away from the body.

   Answer A is incorrect. It is advisable to unhook the negative battery cable and wait about 10 minutes before working on or around the airbag components.

   Answer B is incorrect. Inflatable airbag components should be stored "face up" in a secure area while performing repairs on an airbag-equipped vehicle.

   Answer C is correct. It is never advisable to alter the airbag indicator in any way.

   Answer D is incorrect. It is advisable to carefully carry inflatable airbag components, holding them away from the body.

8. Any of these are methods to verify that the evaporator core is leaking EXCEPT:

TASK B.2.8

   A. Using an electronic leak detector at the A/C duct outlets
   B. Using an ultrasonic leak detector at the HVAC case drain tube
   C. Injecting dye into the system and using a black light to inspect the substance exiting the HVAC case drain tube
   D. Using a refrigerant identifier at the low-side fitting

   Answer A is incorrect. This method could be used if the evaporator core has a severe leak.

   Answer B is incorrect. If the evaporator core had a very large leak, this method could be used.

   Answer C is incorrect. Once dye has been added, then an evaporator core leak will cause the HVAC case drain tube to glow bright yellow as it drains the water.

   Answer D is correct. A refrigerant identifier is used to test the A/C system for impurities.

9. A vehicle with rear heat and air has a rear heater that does not get warm enough. Technician A says that a crimped heater pipe could be the cause. Technician B says that a partially blocked rear heater core could be the cause. Who is correct?

TASK A.20

   A. A only
   B. B only
   C. Both A and B
   D. Neither A nor B

   Answer A is incorrect. Technician B is also correct.

   Answer B is incorrect. Technician A is also correct.

   Answer C is correct. Both Technicians are correct. A crimped heater pipe could reduce coolant flow to the rear heater core, which would limit heater output temperature. A partially blocked rear heater core could also cause cooler rear heater output temperature.

   Answer D is incorrect. Both Technicians are correct.

TASK A.4, A.8

10. An R-134a A/C system is being recharged. Technician A says that an A/C system can be accurately charged by monitoring pressure and temperature. Technician B says that the A/C system should be treated with refrigerant dye any time that work has been performed so that future leak testing will be possible. Who is correct?

A. A only
B. B only
C. Both A and B
D. Neither A nor B

Answer A is incorrect. Accurate recharging of an A/C system requires the use of some type of scale in order to weigh the refrigerant.

Answer B is correct. Only Technician B is correct. It is a good practice to use refrigerant dye on any system that is worked on in order to leak test the system.

Answer C is incorrect. Only Technician B is correct.

Answer D is incorrect. Technician B is correct.

TASK C.1.1

11. The HVAC fuse blows each time the A/C switch is turned on. Technician A says that an open low-side switch could be the cause. Technician B says that a shorted A/C clutch relay coil could be the cause. Who is correct?

A. A only
B. B only
C. Both A and B
D. Neither A nor B

Answer A is incorrect. An open circuit prevents any current from flowing past the open point. Fuses typically blow when too much current flows in the circuit.

Answer B is correct. Only Technician B is correct. A short in the A/C clutch relay coil could cause current to be very high and blow the HVAC fuse. A short to ground at some other point in the circuit could also cause the fuse to blow.

Answer C is incorrect. Only Technician B is correct.

Answer D is incorrect. Technician B is correct.

TASK A.12

12. A vehicle with an automatic temperature control (ATC) system is being diagnosed. Technician A says that the ATC trouble codes can be retrieved by depressing a sequence of buttons on the ATC control head. Technician B says that the trouble code reveals which area needs to be repaired in an ATC system. Who is correct?

A. A only
B. B only
C. Both A and B
D. Neither A nor B

Answer A is incorrect. Technician B is also correct.

Answer B is incorrect. Technician A is also correct.

Answer C is correct. Both Technicians are correct. Many ATC systems are designed to display ATC trouble codes by depressing a sequence of buttons on the ATC control head. The trouble codes reveals which area needs to be diagnosed so the technician can then investigate by performing extensive tests on the system to find a root cause of the problem.

Answer D is incorrect. Both Technicians are correct.

13. Any of these conditions can cause elevated high-side pressures EXCEPT:

TASK A.1

A. Refrigerant overcharge
B. Restricted airflow to the condenser
C. Poor airflow across the evaporator
D. A slipping fan clutch

Answer A is incorrect. Having too much refrigerant will always cause elevated high-side pressures and poor A/C performance.

Answer B is incorrect. An airflow restriction at the condenser will cause elevated high-side pressures due to the lack of heat transfer in the condenser.

Answer C is correct. An airflow problem at the evaporator will not cause elevated high-side pressures. This problem would cause reduced airflow at the HVAC ducts and grids.

Answer D is incorrect. A slipping fan clutch would cause elevated high-side pressures when the vehicle is traveling at low speeds or sitting in traffic.

14. Which of these oils is the standard oil used by OEMs in (nonhybrid) R-134a systems?

TASK A.9

A. Mineral oil
B. 10W30
C. Ester oil
D. Polyalkylene glycol (PAG)

Answer A is incorrect. Mineral oil is used in R-12 systems but it is too heavy to be used in R-134a systems.

Answer B is incorrect. 10W30 is a multi-viscosity lubricant used in many engines.

Answer C is incorrect. Ester oil is an oil that is sometimes used in vehicles that have been retrofitted from R-12 to R-134a.

Answer D is correct. PAG oil is a synthetic refrigerant oil that is used in (nonhybrid) factory R-134a systems.

15. The heating system is being diagnosed for poor heating performance. Technician A says the problem could be a stuck open engine thermostat. Technician B says the problem could be an inoperative cooling fan. Who is right?

TASK A.20

A. A only
B. B only
C. Both A and B
D. Neither A nor B

Answer A is correct. Only Technician A is correct. A stuck open engine thermostat will cause the engine to not maintain normal operating temperature.

Answer B is incorrect. An inoperative cooling fan would cause the engine to run hotter than normal at slow speeds and while sitting still.

Answer C is incorrect. Only Technician A is correct.

Answer D is incorrect. Technician A is correct.

TASK A.4

16. An A/C system is being diagnosed for a leak. Technician A carefully checks all system connections and says the presence of oil around the fitting of an air conditioning line or hose is an indication of a possible refrigerant leak. Technician B uses a handheld electronic leak detector and moves the hose slowly along all components of the refrigerant system. Who is correct?

    A. A only
    B. B only
    C. Both A and B
    D. Neither A nor B

    Answer A is incorrect. Technician B is also correct.

    Answer B is incorrect. Technician A is also correct.

    Answer C is correct. Both Technicians are correct. The presence of oily residue around A/C components is usually a sign of a refrigerant leak. An electronic leak detector is often a useful tool that can be used to find A/C leaks.

    Answer D is incorrect. Both Technicians are correct.

TASK B.2.9

17. The fins and air passages of an evaporator are heavily clogged and the airflow has been greatly reduced. Technician A says the evaporator core must be removed from the case for proper cleaning. Technician B says this condition can cause extra load on the blower motor. Who is correct?

    A. A only
    B. B only
    C. Both A and B
    D. Neither A nor B

    Answer A is incorrect. Technician B is also correct.

    Answer B is incorrect. Technician A is also correct.

    Answer C is correct. Both Technicians are correct. An extremely clogged evaporator core can only be cleaned by disassembling the case and removing the evaporator core to thoroughly clean. A clogged evaporator core would cause extra mechanical load on the blower motor and could lead to early failure.

    Answer D is incorrect. Both Technicians are correct.

TASK B.2.6

18. Technician A says a restricted orifice tube could be caused by debris from a failing compressor. Technician B says an orifice tube restriction would cause a low-side pressure that is considerably higher than specified. Who is correct?

    A. A only
    B. B only
    C. Both A and B
    D. Neither A nor B

    Answer A is correct. Only Technician A is correct. The metal screen on the orifice tube is often restricted from debris that has been pumped through the system. The two common places that the debris comes from are the compressor and the accumulator/drier.

    Answer B is incorrect. A restricted orifice tube would cause the low-side pressure to be lower than normal and would also cause reduced cooling performance.

    Answer C is incorrect. Only Technician A is correct.

    Answer D is incorrect. Technician A is correct.

19. Technician A says the desiccant in a receiver/drier absorbs moisture and it can become quickly contaminated if exposed to atmosphere. Technician B says orifice tube systems use an accumulator to store excess refrigerant and to filter and dry the refrigerant. Who is correct?

TASK B.2.4

A. A only
B. B only
C. Both A and B
D. Neither A nor B

Answer A is incorrect. Technician B is also correct.

Answer B is incorrect. Technician A is also correct.

Answer C is correct. Both Technicians are correct. The desiccant in drier devices does absorb moisture. Care should be taken when installing new drier devices to make sure that they are installed last in the repair process. After installation, the system should be evacuated for at least 30 minutes to remove all possible moisture from the system. A/C systems with orifice tubes do use an accumulator drier.

Answer D is incorrect. Both Technicians are correct.

20. Any of these statements about refrigerant lines are true EXCEPT:

TASK B.2.1

A. Suction lines are located between the outlet side of the evaporator and the inlet side or suction side of the compressor.
B. Suction lines carry the low-pressure, low-temperature refrigerant vapor to the compressor where it again is moved through the system.
C. Suction lines can be distinguished from the discharge lines by touch: They are hot to the touch when the system is operating.
D. The suction line is larger in diameter than the liquid line because refrigerant in a vapor state takes up more room than refrigerant in a liquid state.

Answer A is incorrect. The A/C line connecting the evaporator core and the compressor is called the suction line. When the system is operating, the refrigerant is a low-pressure gas.

Answer B is incorrect. The A/C suction does carry low-pressure gas from the evaporator to the compressor.

Answer C is correct. Suction lines are cold to the touch when the system is operating.

Answer D is incorrect. The suction line is always larger than the liquid line.

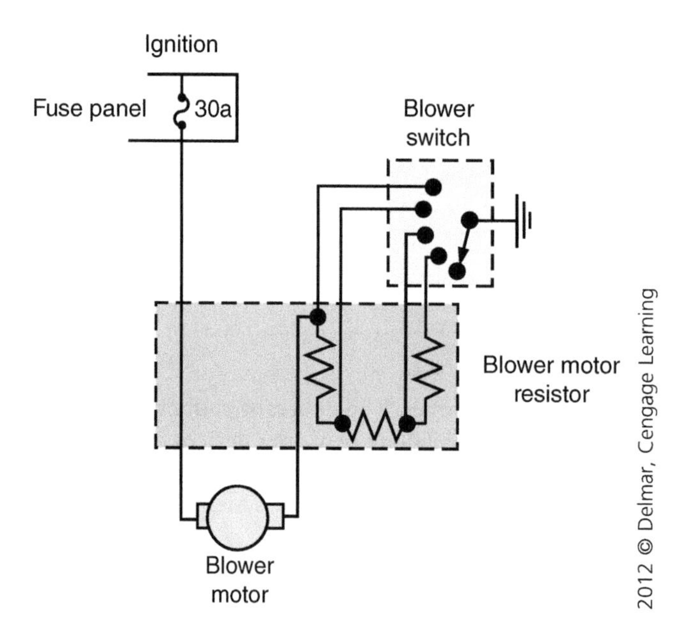

TASK C.1.2

21. In the figure above, the blower motor only works on high speed. Technician A says that a faulty ground at the blower switch could be the cause. Technician B says that an open blower resistor could be the cause. Who is correct?

A. A only
B. B only
C. Both A and B
D. Neither A nor B

Answer A is incorrect. If the switch ground was faulty, then the blower would not work on any speed.

Answer B is correct. Only Technician B is correct. An open blower resistor could cause the blower to only work on high speed.

Answer C is incorrect. Only Technician B is correct.

Answer D is incorrect. Technician B is correct.

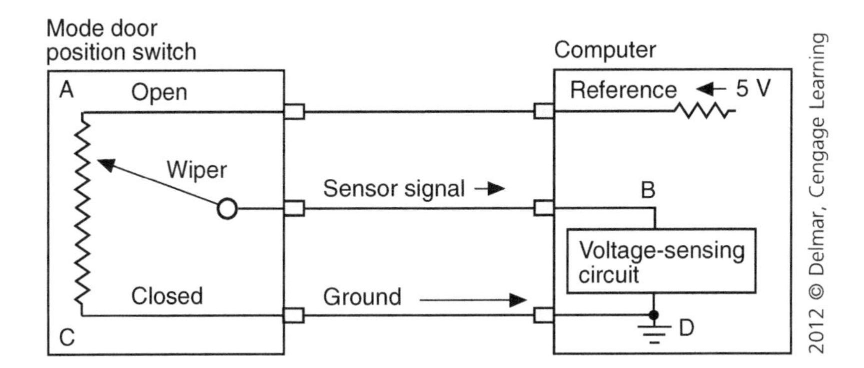

22. Referring to the figure above, the mode door actuator is inoperative. Technician A says that the computer senses the position of the mode door by monitoring the reference voltage. Technician B says that the computer provides the ground for the mode door position switch. Who is correct?

TASK C.1.7

A. A only
B. B only
C. Both A and B
D. Neither A nor B

Answer A is incorrect. The computer monitors the position of the mode door by monitoring the signal voltage, which is the middle wire of the potentiometer.

Answer B is correct. Only Technician B is correct. The computer provides the ground for the mode door position switch. This device is known as a potentiometer, which is commonly used as a position sensor.

Answer C is incorrect. Only Technician B is correct.

Answer D is incorrect. Technician B is correct.

TASK B.1.5

23. Referring to the figure above, Technician A says that the old compressor oil should be drained and measured when replacing an A/C compressor. Technician B says that the old oil should be added to the new compressor. Who is correct?

A. A only
B. B only
C. Both A and B
D. Neither A nor B

Answer A is correct. Only Technician A is correct. The oil should be drained from the old compressor whenever a replacement compressor is to be installed. If more than two ounces of oil were drained from the old compressor, then that amount should be added to the new compressor. If less than two ounces of oil were drained from the old compressor, then two ounces of new oil should be added to the new compressor.

Answer B is incorrect. Used oil should never be put back into an A/C component or system. New oil should be added to the new compressor (unless otherwise stated by the compressor manufacturer).

Answer C is incorrect. Only Technician A is correct.

Answer D is incorrect. Technician A is correct.

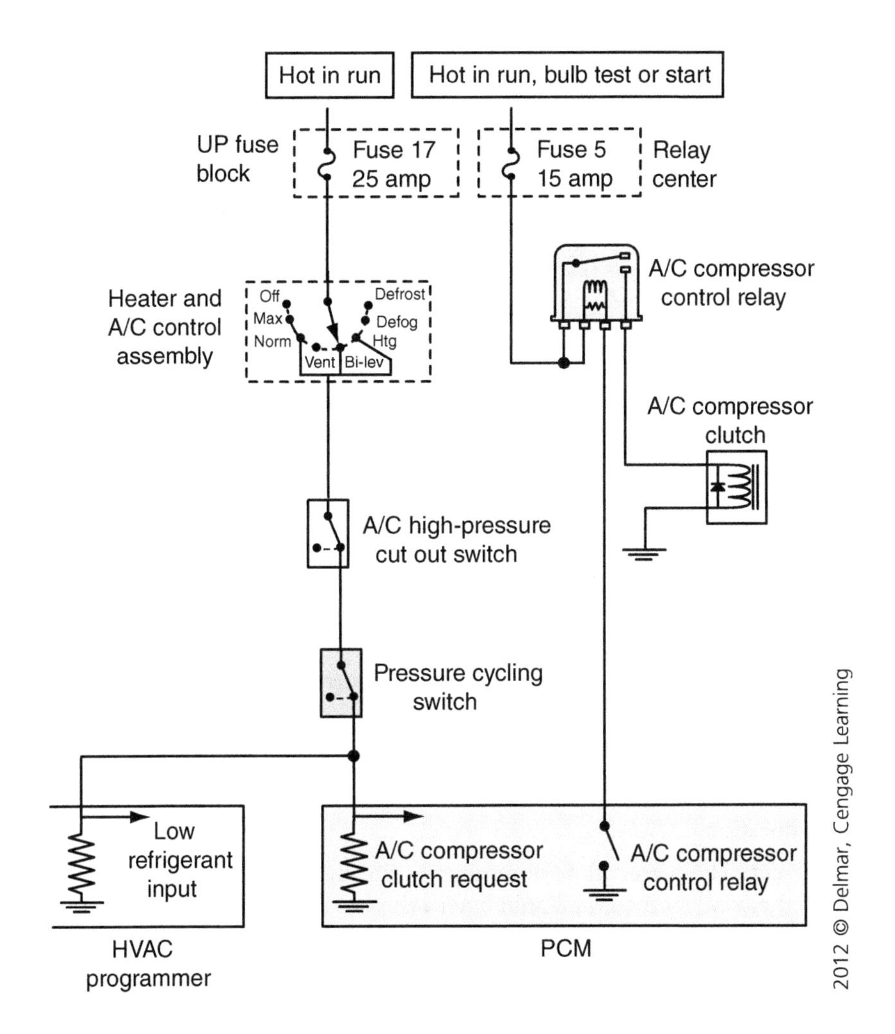

24. The A/C clutch in the figure above will not engage. Which of these would be the LEAST LIKELY cause?

A. A blown fuse 17
B. A stuck closed A/C compressor control relay
C. A faulty PCM relay driver
D. A faulty pressure cycling switch

Answer A is incorrect. A blown fuse 17 would cause the A/C clutch to not engage. Fuse 17 supplies the signal voltage that is routed through the HVAC control head and switches.

Answer B is correct. The compressor would run constantly if the A/C compressor control relay was stuck closed.

Answer C is incorrect. The relay would never turn on if the PCM relay driver was faulty.

Answer D is incorrect. A faulty pressure cycling switch would prevent the A/C request signal from reaching the PCM.

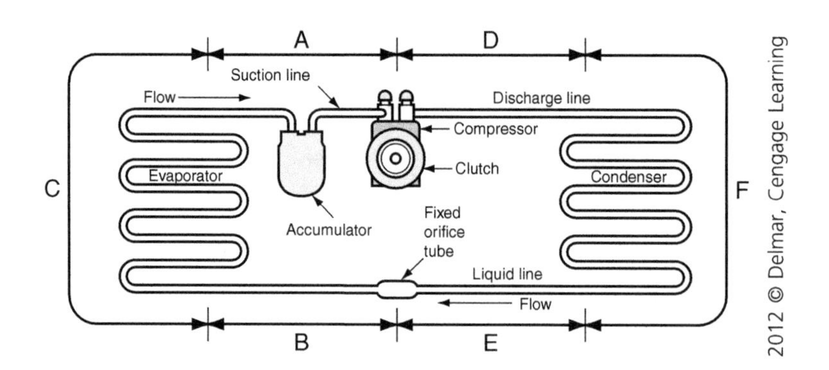

TASK A.1

25. Referring to the figure above, list the correct flow of refrigerant from the compressor exit point back to the compressor inlet point.

A. A to C to B to E to F to D
B. D to A to C to B to E to F
C. B to C to A to D to F to E
D. D to F to E to B to C to A

Answer A is incorrect. This path is opposite of what the correct path would be.

Answer B is incorrect. The flow does not go from the discharge line directly into the suction line. The direction is also opposite of what the correct path would be.

Answer C is incorrect. This path begins at the metering device, not at the compressor as described in the question.

Answer D is correct. The correct flow of refrigerant through the system is through the discharge line to the condenser to the liquid line to the metering device to the evaporator core to the accumulator to the compressor inlet.

TASK B.1.1

26. A vehicle with a faulty A/C pressure sensor is being diagnosed. Technician A says that this device is used to provide A/C pressure feedback to a control module. Technician B says that this device is typically mounted in the low side of the A/C system. Who is correct?

A. A only
B. B only
C. Both A and B
D. Neither A nor B

Answer A is correct. Only Technician A is correct. A/C pressure sensors are often used on the high side of A/C systems to provide a voltage signal to a control module. This varying voltage is a more accurate method of controlling the A/C system. The range of the voltage is approximately 1 to 3 volts during operation.

Answer B is incorrect. A/C pressure sensors are most often mounted somewhere in the high side of the system.

Answer C is incorrect. Only Technician A is correct.

Answer D is incorrect. Technician A is correct.

27. The most likely problem that would result from an open blower resistor would be:

TASK C.1.2

    A. The blower would work on all lower speeds but not on high speed.
    B. The blower would not work on any speed.
    C. The blower would only work on high speed.
    D. The blower fuse would blow each time the blower is turned on at any speed.

Answer A is incorrect. This fault would typically be caused by a bad blower relay or a bad blower switch.

Answer B is incorrect. This fault would typically be caused by a blown fuse or a bad control head not feeding any current to the blower circuit.

Answer C is correct. An open blower resistor would typically cause all of the lower blower speeds to be inoperative, but high speed would still operate since the current bypasses the blower resistor on high speed.

Answer D is incorrect. A short to ground would typically be the cause of the fuse blowing each time the blower is turned on at any speed.

28. An A/C compressor cycles off when the high-side pressure builds. Technician A says that the high-pressure cutoff switch is opening the circuit to the compressor clutch. Technician B says that the pressure release valve is opening the circuit to the compressor clutch. Who is correct?

TASK B.2.10

    A. A only
    B. B only
    C. Both A and B
    D. Neither A nor B

Answer A is correct. Only Technician A is correct. A high-pressure cutoff switch opens its contacts when the pressure exceeds 450 psi.

Answer B is incorrect. The pressure release valve is a mechanical device that releases excess A/C pressure to prevent a component from being ruptured due to high pressure.

Answer C is incorrect. Only Technician A is correct.

Answer D is incorrect. Technician A is correct.

29. A performance test is being performed on the A/C system. Technician A says that the A/C system should be turned on normal A/C and the blower set to low speed. Technician B says that the engine should be at idle during this test. Who is correct?

TASK A.1

    A. A only
    B. B only
    C. Both A and B
    D. Neither A nor B

Answer A is incorrect. The A/C controls should be set on maximum A/C and the blower should be set to high speed during a performance test.

Answer B is incorrect. The engine RPM should be increased to approximately 1500 during an A/C performance test. This allows the compressor to pump more refrigerant to assist with removing the heat from the cabin area.

Answer C is incorrect. Neither Technician is correct.

Answer D is correct. Neither Technician is correct.

TASK B.1.1, B.1.2

30. The A/C compressor will not engage when the A/C is turned on. The static refrigerant pressure is 75 psi and the outside air temperature is 72°F. Technician A says that a poor connection at the pressure cycling switch could be the cause. Technician B says that a faulty A/C clutch coil could be the cause. Who is correct?

    A. A only
    B. B only
    C. Both A and B
    D. Neither A nor B

    Answer A is incorrect. Technician B is also correct.

    Answer B is incorrect. Technician A is also correct.

    Answer C is correct. Both Technicians are correct. A bad connection at the pressure cycling switch would interrupt the A/C request signal to the ECM, which would cause the compressor to not engage. A faulty A/C clutch coil could also cause this problem because it could not produce a magnetic field when energized.

    Answer D is incorrect. Both Technicians are correct.

TASK C.2.1, C.2.4

31. The vacuum supply hose has been broken at the vacuum check valve. Technician A says that the temperature door will be stuck in the full hot position. Technician B says the defrost door will be stuck in the defrost position. Who is correct?

    A. A only
    B. B only
    C. Both A and B
    D. Neither A nor B

    Answer A is incorrect. The temperature door is never activated by a vacuum device. The temperature door needs to have many variables that a vacuum device cannot provide.

    Answer B is correct. Only Technician B is correct. If vacuum is lost, most HVAC systems will default to the defrost setting by design. This feature is in place for safety purposes.

    Answer C is incorrect. Only Technician B is correct.

    Answer D is incorrect. Technician B is correct.

TASK A.20, A.25, A.28

32. A vehicle is in the shop with the complaint of the heater not being hot enough. The coolant level has been checked and is found to be full. Technician A says that the thermostat should be tested for proper operation. Technician B says that the heater control valve could be stuck open. Who is correct?

    A. A only
    B. B only
    C. Both A and B
    D. Neither A nor B

    Answer A is correct. Only Technician A is correct. The thermostat should be checked for proper operation any time the heater performance is in question. The technician can watch the temperature gauge or use a scan tool to monitor engine temperature.

    Answer B is incorrect. A stuck open heater control valve will not cause a heater performance problem. It could potentially cause an A/C performance problem.

    Answer C is incorrect. Only Technician A is correct.

    Answer D is incorrect. Technician A is correct.

33. A vehicle is in the shop for an overheating problem. During the inspection the technician finds that the upper radiator hose collapses after the engine cools down but moves back to normal when the radiator cap is removed. The most likely cause for this problem is?

TASK A.24

A. Faulty water pump
B. Malfunctioning cooling fan
C. Faulty radiator cap
D. Faulty thermostat

Answer A is incorrect. A faulty water pump could fail by leaking or by not pumping enough coolant, but would not cause the conditions described in this question.

Answer B is incorrect. A malfunctioning cooling fan could cause overheating at slow speeds or while parked, but would not cause the conditions described in this question.

Answer C is correct. The vacuum valve in the radiator cap is not allowing the coolant to return to the radiator when the temperature cools down again. This problem is signaled by the upper radiator hose collapsing.

Answer D is incorrect. A faulty thermostat can stick open and cause overcooling as well as sticking closed and cause overheating, but it would not cause the conditions described in this question.

34. What function does the A/C clutch coil diode perform for the A/C system?

TASK C.1.3

A. Prevents the voltage spike from damaging other components
B. Assists the clutch coil in creating magnetism
C. Limits current flow to prevent the A/C fuse from overheating
D. Connects the clutch coil to ground

Answer A is correct. The A/C clutch diode is wired in parallel with the coil and works to route the inductive voltage spike safely back into the coil.

Answer B is incorrect. The A/C clutch diode does not assist in creating magnetism. The magnetic field is produced by the coil when it is energized.

Answer C is incorrect. The A/C clutch diode does not limit current flow.

Answer D is incorrect. The A/C clutch coil does not connect the coil to ground. The coil is either case grounded or it has a wire connecting it to ground.

35. Any of these are acceptable results from an A/C performance test EXCEPT:

TASK A.1, A.3

A. Duct temperatures ranging from 40°F to 50°F (4.4°C–10°C).
B. High-side pressure at approximately two to three times the ambient temperature
C. Low-side pressure at approximately 50 to 70 psi (343.8 kPa to 482.6 kPa)
D. Suction line is cold to the touch

Answer A is incorrect. The duct temperatures should be in the 40°F to 50°F (4.4°C–10°C). range during an A/C performance test.

Answer B is incorrect. The high-side pressure should be approximately two to three times the ambient temperature during an A/C performance test.

Answer C is correct. The low-side pressure should be about 25 to 35 psi (172.4 kPa to 241.3 kPa) during an A/C performance test.

Answer D is incorrect. The suction line should be cold to the touch during an A/C performance test because it contains low-pressure cold refrigerant vapor.

TASK C.1.4

36. Which device is used as the input to cause the A/C clutch to disengage under heavy engine loads?

    A. Coolant temperature sensor
    B. Throttle position sensor
    C. Oxygen sensor
    D. Power steering switch

Answer A is incorrect. The coolant temperature sensor is a variable resistor that provides engine temperature data to the ECM.

Answer B is correct. The throttle position sensor is a variable resistor that provides throttle angle data to the ECM. When this sensor shows heavy throttle angles, the ECM will de-energize the A/C to save engine horsepower.

Answer C is incorrect. The oxygen sensor provides fuel control data to the ECM by signaling the amount of oxygen in the exhaust after the combustion process has taken place.

Answer D is incorrect. The power steering switch sends power steering information to the ECM. This switch opens when high power steering pressure is present, such as during a parking maneuver.

TASK D.2.4

37. Vacuum actuators can be used for any of these duct box doors EXCEPT:

    A. Temperature door
    B. Defrost mode door
    C. Fresh air door
    D. Heat mode door

Answer A is correct. The temperature door needs many different angles that cannot be provided with a vacuum actuator. Temperature doors are typically moved by cables, linkage, or electric motors.

Answer B is incorrect. The defrost mode door is activated by a vacuum actuator on many late-model vehicles.

Answer C is incorrect. The fresh air door is activated by a vacuum actuator on many late-model vehicles.

Answer D is incorrect. The heat mode door is activated by a vacuum actuator on many late-model vehicles.

TASK C.1.5

38. The thermal limiter is being diagnosed on a late-model vehicle. Technician A says that this device can be checked with a scan tool by using a function test of the cooling fan. Technician B says that this device can be checked with an ohmmeter when the power is turned off. Who is correct?

    A. A only
    B. B only
    C. Both A and B
    D. Neither A nor B

Answer A is incorrect. The scan tool cooling fan output test would not test the thermal limiter.

Answer B is correct. Only Technician B is correct. The thermal limiter is an inline thermal circuit protection device that can be tested with an ohmmeter with the circuit turned off.

Answer C is incorrect. Only Technician B is correct.

Answer D is incorrect. Technician B is correct.

39. Technician A says that the evaporator temperature switch is a thermostatic device that opens when the evaporator core gets near the freezing level. Technician B says that the evaporator temperature sensor is a thermistor that varies as the temperature changes in the evaporator core. Who is correct?

TASK B.1.2

A. A only
B. B only
C. Both A and B
D. Neither A nor B

Answer A is incorrect. Technician B is also correct.

Answer B is incorrect. Technician A is also correct.

Answer C is correct. Both Technicians are correct. The evaporator temperature switch opens to prevent evaporator freeze-up and the evaporator temperature sensor is a variable temperature resistor that also helps prevent evaporator freeze-up.

Answer D is incorrect. Both Technicians are correct.

40. A vehicle with automatic temperature control (ATC) is malfunctioning. A scan tool is connected and a code is retrieved from the system concerning the cabin temperature sensor. Technician A says that the sensor is located near the engine air filter. Technician B says the sensor can be checked with an ohmmeter to see if it has the correct resistance. Who is correct?

TASK C.3.1

A. A only
B. B only
C. Both A and B
D. Neither A nor B

Answer A is incorrect. The cabin air temperature sensor is located inside the cabin area to sense cabin temperature.

Answer B is correct. Only Technician B is correct. An ohmmeter could be used to test the cabin temperature sensor resistance. This sensor is a thermistor that should vary its resistance as the temperature is varied.

Answer C is incorrect. Only Technician B is correct.

Answer D is incorrect. Technician B is correct.

41. Any of these methods are used to move the temperature door EXCEPT:

TASK C.3.6

A. Electronic actuators
B. Vacuum actuators
C. Cables
D. Linkages

Answer A is incorrect. Electronic actuators are commonly used to move the temperature blend door.

Answer B is correct. Vacuum is never used as a method to move the temperature door because of the limited possible positions that these devices possess.

Answer C is incorrect. Manual controls connected to Bowden cables are commonly used to move the temperature blend door.

Answer D is incorrect. Manual controls and linkages are sometimes used to move the temperature blend door.

TASK B.1.4

42. An A/C compressor is noisy when the clutch is engaged but the noise stops when the clutch is disengaged. Which of these is the most likely cause?

 A. Faulty compressor clutch pulley bearing
 B. Low refrigerant charge
 C. Internal compressor damage
 D. Overtightened drive belt

Answer A is incorrect. A pulley bearing will make noise when the compressor is turned off and then the noise will stop when the compressor is turned on.

Answer B is incorrect. A low refrigerant charge will typically cause poor A/C performance as well as a short cycling compressor rate.

Answer C is correct. An internal compressor problem will cause noise when the compressor is engaged and the noise will stop when the compressor is not on.

Answer D is incorrect. An overtightened A/C drive belt can cause bearing fatigue in the driven accessories. This noise would not stop when the compressor is disengaged.

TASK A.17

43. A refillable A/C refrigerant cylinder is considered full when it reaches what capacity by weight?

 A. 50 percent
 B. 60 percent
 C. 70 percent
 D. 80 percent

Answer A is incorrect. This level is too low.

Answer B is correct. The tank must not be filled beyond 60 percent of the gross weight rating. Some room must be left in the cylinder to make room if the temperature is raised.

Answer C is incorrect. This level is too high.

Answer D is incorrect. This level is too high.

TASK A.5

44. The refrigerant is being recovered from an A/C system. Five minutes after the recovery process is complete, the low-side pressure loses the vacuum and the pressure rises above zero. This condition indicates:

 A. There is still some refrigerant in the system.
 B. There is excessive oil in the refrigerant system.
 C. The refrigerant system is leaking.
 D. There is excessive moisture in the refrigerant system.

Answer A is correct. If the low-side gauge rises above zero after a short period, then there is likely still refrigerant present in the A/C system and the system must be recovered again.

Answer B is incorrect. Excessive oil in the system could cause slightly elevated operating pressures, but not the condition explained in the question.

Answer C is incorrect. A leaking refrigerant system could cause the system to lose its vacuum, but it would not cause the pressure to rise above zero.

Answer D is incorrect. Excessive moisture in the system can cause long-term problems by producing acid. In addition, moisture can cause the system to develop ice near the metering device, but it will not cause the condition explained in the question.

45. A cycling clutch orifice tube (CCOT) A/C system is operating at 82°F (27.8°C) ambient temperature, the compressor clutch cycles several times per minute, and the suction line is warm. The high-side gauge shows lower than normal pressures. The most likely cause of this problem could be:

TASK A.3, A.2

A. A low refrigerant charge
B. A flooded evaporator
C. A restricted accumulator
D. An overcharge of refrigerant

Answer A is correct. A low refrigerant charge would cause short cycling at the compressor as well as low system pressures.

Answer B is incorrect. If the evaporator was flooded, then the suction line would be frosted due to the refrigerant still taking on heat.

Answer C is incorrect. A restricted accumulator would cause the suction line to be frosted due to the restriction.

Answer D is incorrect. A refrigerant overcharge would cause the high-side pressure to be higher than normal.

46. A vehicle has a problem of a stalling engine when the steering wheel is turned to full lock. This problem happens only when the air conditioning is on. Technician A says that the vehicle may have a bad power steering pressure switch. Technician B says this could occur if the power steering belt is loose. Who is correct?

TASK C.1.5

A. A only
B. B only
C. Both A and B
D. Neither A nor B

Answer A is correct. Only Technician A is correct. The power steering switch sends a signal to the ECM to de-energize the A/C system when high power steering pressure is present.

Answer B is incorrect. A loose power steering belt would cause belt noise during high power steering system load. It would not cause the engine to stall.

Answer C is incorrect. Only Technician A is correct.

Answer D is incorrect. Technician A is correct.

47. An air distribution vacuum actuator circuit is being tested using a vacuum gauge. The gauge is connected into the vacuum line going to the engine, and a zero vacuum reading is displayed. This would indicate:

TASK C.2.1

A. The HVAC vacuum switching valve is faulty.
B. The line to the engine is plugged, disconnected, or kinked.
C. The vacuum actuator is defective.
D. The actuator should be operational.

Answer A is incorrect. The test described above would not test the HVAC vacuum valve.

Answer B is correct. If vacuum is not present in this hose, then the line is either plugged, disconnected, or kinked. The gauge should be reading approximately 18 in. Hg at idle.

Answer C is incorrect. The test described in this question would not test the vacuum actuator.

Answer D is incorrect. The vacuum system would not be operational due to the lack of supply vacuum.

TASK A.1, A.2

48. The high-side pressure on a system with a cycling clutch and pressure cycling switch is 245 psi, the low-side pressure hovers around 28 psi, and the ambient temperature is 98°F. What do these gauge readings indicate?

A. The system is undercharged.
B. The system is overcharged.
C. The system is normal.
D. The evaporator pressure regulator is bad.

Answer A is incorrect. An undercharged A/C system would produce much lower pressures than those present in this question.

Answer B is incorrect. An overcharged A/C system would produce much higher pressures than those present in this question.

Answer C is correct. The pressures in this question would be very good for the extremely hot day described in this question.

Answer D is incorrect. The pressures are in the normal range. Evaporator pressure regulators were used on A/C systems over 25 years ago.

TASK C.1.2

49. A vehicle is being diagnosed for the cause of a blown fuse in the blower motor circuit. Technician A says a short to ground in the circuit caused the fuse to blow. Technician B says an open field winding in the fan motor could have caused the fuse to blow. Who is correct?

A. A only
B. B only
C. Both A and B
D. Neither A nor B

Answer A is correct. Only Technician A is correct. A short to ground that happens before the intended load will usually blow the circuit protection device.

Answer B is incorrect. An open circuit would cause an interruption in current flow, which would never cause a blown fuse.

Answer C is incorrect. Only Technician A is correct.

Answer D is incorrect. Technician A is correct.

TASK C.3.4

50. While testing a compressor clutch, a fused jumper wire is used to bypass the load side of the A/C clutch relay. This does not cause the compressor clutch to engage. The LEAST LIKELY cause of this would be:

A. An open compressor clutch coil
B. A shorted compressor clutch coil
C. An open compressor clutch coil ground circuit
D. An open relay power feed circuit

Answer A is incorrect. An open compressor clutch coil would cause the condition described in this question.

Answer B is correct. A shorted compressor clutch coil would cause the fused jumper wire to burn the fuse. It would not cause the condition described in the question.

Answer C is incorrect. An open compressor clutch coil ground circuit would cause the condition described in this question.

Answer D is incorrect. An open compressor clutch coil in the relay power circuit would cause the condition described in this question.

PREPARATION EXAM 2—ANSWER KEY

1. B
2. D
3. A
4. C
5. D
6. C
7. B
8. D
9. A
10. A
11. B
12. A
13. D
14. C
15. D
16. B
17. C
18. B
19. A
20. A
21. C
22. C
23. D
24. C
25. A
26. C
27. C
28. C
29. A
30. C
31. D
32. D
33. B
34. C
35. B
36. D
37. D
38. B
39. A
40. A
41. D
42. C
43. A
44. D
45. C
46. C
47. A
48. B
49. B
50. D

PREPARATION EXAM 2—EXPLANATIONS

1. What would be the most likely result of a missing A/C compressor clutch diode?

TASK C.1.3

A. Inoperative A/C compressor
B. Damage to the ECM
C. Compressor will not disengage
D. Rear defogger will not engage

Answer A is incorrect. The compressor would still operate but it does not have clamping protection when the coil is de-energized.

Answer B is correct. The clamping diode at the A/C compressor clutch suppresses the voltage spike that is created when the compressor is de-energized. If this diode was missing, then dangerous voltage could potentially reach the ECM.

Answer C is incorrect. The compressor would still engage if the diode was missing.

Answer D is incorrect. The A/C clutch diode would not likely have any effect on the rear defogger.

TASK B.2.11

2. Technician A says that the service fittings for R-134a are the same as R-12 service fittings. Technician B says that the low-side service fitting is a 16 mm quick-connect style and the high-side fitting is a 13 mm quick-connect style. Who is correct?

   A. A only
   B. B only
   C. Both A and B
   D. Neither A nor B

   Answer A is incorrect. R-134a fittings are 13 mm and 16 mm quick-connect styles and R-12 fittings are 3/8 and 7/16 threaded styles.

   Answer B is incorrect. The low-side fitting is a 13 mm quick-connect style and the high-side fitting is a 16 mm quick-connect style.

   Answer C is incorrect. Neither Technician is correct.

   Answer D is correct. Neither Technician is correct. R-134a fittings are different from R-12 fittings. The low-side fittings for R-134a are 13 mm quick-connect style. The high-side fittings for R-134a are 16 mm quick-connect style.

TASK C.2.3

3. An inoperative cable-controlled heater control valve is being diagnosed. The control moves freely, but the valve does not respond. Technician A says the cable housing clamp may be loose at the control head (panel) end. Technician B says the cable may be rusted in the housing. Who is correct?

   A. A only
   B. B only
   C. Both A and B
   D. Neither A nor B

   Answer A is correct. Only Technician A is correct. A loose cable housing clamp could cause the heater control valve to not move when the control was moved.

   Answer B is incorrect. The control would not move freely if the cable was rusted in the housing.

   Answer C is incorrect. Only Technician A is correct.

   Answer D is incorrect. Technician A is correct.

TASK C.3.11

4. Technician A says that the scan tool receives data from the vehicle by communicating on the data bus network. Technician B says that if one of the two data bus wires becomes broken, then the network can still communicate on the remaining good wire but will set a trouble code. Who is correct?

   A. A only
   B. B only
   C. Both A and B
   D. Neither A nor B

   Answer A is incorrect. Technician B is also correct.

   Answer B is incorrect. Technician A is also correct.

   Answer C is correct. Both Technicians are correct. The scan tool does receive data over the data network. In addition, the data network can typically still have limited communication if one of the wires breaks. A trouble code should set and alert the driver with a warning light.

   Answer D is incorrect. Both Technicians are correct.

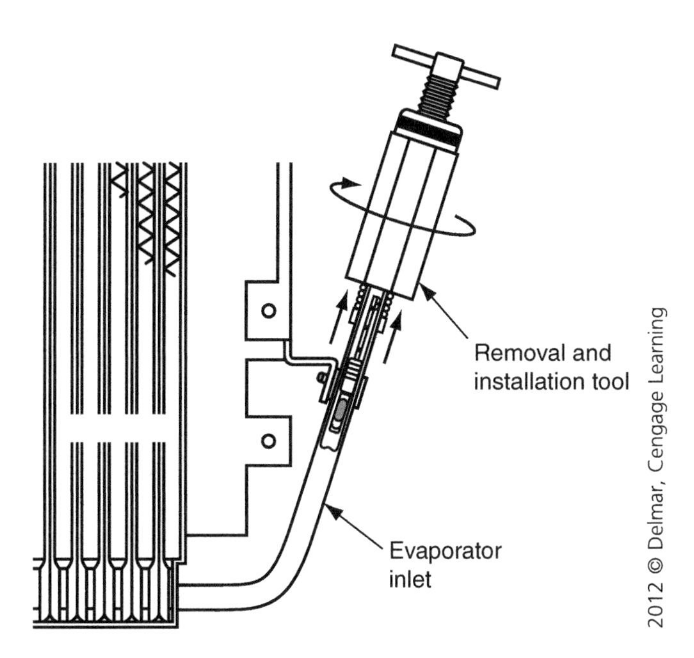

5. The tool in the figure above is used to remove and install which A/C component?

A. The evaporator core
B. The heater core
C. The TXV
D. The orifice tube

Answer A is incorrect. The orifice tube remover is not used to perform work on the evaporator core.

Answer B is incorrect. The orifice tube remover is not used to perform work on the heater core.

Answer C is incorrect. The orifice tube remover is not used to service a TXV.

Answer D is correct. The tool is used to remove the orifice tube from the liquid line or evaporator core inlet, depending on the manufacturer.

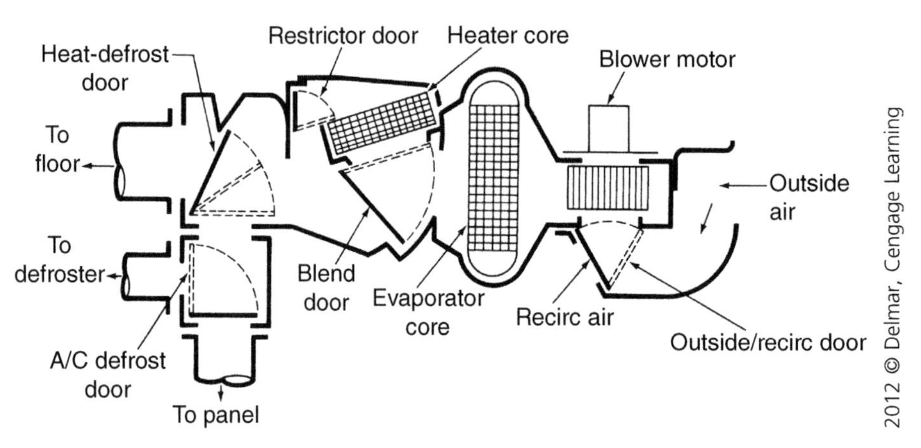

TASK D.2.6

6. Referring to the figure above, the blend door became jammed in the middle of the travel area. What would be the most likely system fault?
   A. The HVAC air would be pulled from outside the car.
   B. The HVAC air would only come out of the heat and defroster grids.
   C. The HVAC air temperature could not be regulated.
   D. The HVAC air would only come out of the vent and heat ducts.

Answer A is incorrect. The blend door controls the temperature of the air. The fresh/recirculated air door controls the origin of the air.

Answer B is incorrect. The mode doors control the location to which air is directed.

Answer C is correct. A jammed blend door would prevent the function of adjusting the air temperature in the HVAC system.

Answer D is incorrect. The mode doors control the location to which air is directed.

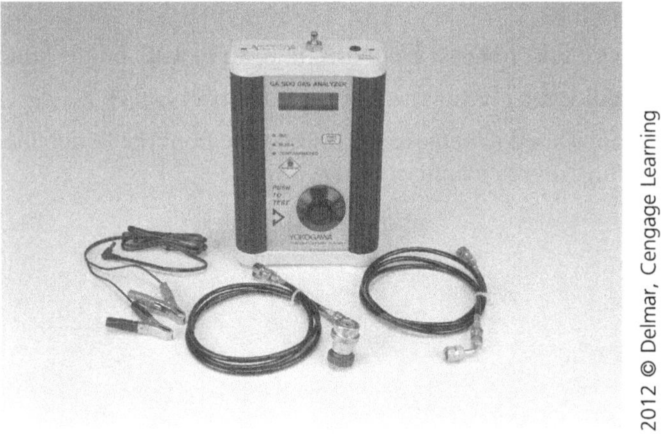

TASK A.5

7. Which A/C-related procedure would the above tool be used for?
   A. Weighing refrigerant during recharging
   B. Testing the refrigerant for impurities
   C. Testing the A/C system performance
   D. Testing the coolant for electrolysis

Answer A is incorrect. A digital scale or an A/C charging machine would be needed to weigh the refrigerant during a recharge process.

Answer B is correct. The refrigerant identifier is used to test for impurities such as air, a flammable substance, or a blended refrigerant.

Answer C is incorrect. A manifold gauge assembly and a thermometer would be needed to test the A/C system performance.

Answer D is incorrect. A digital voltmeter is needed to test for coolant electrolysis.

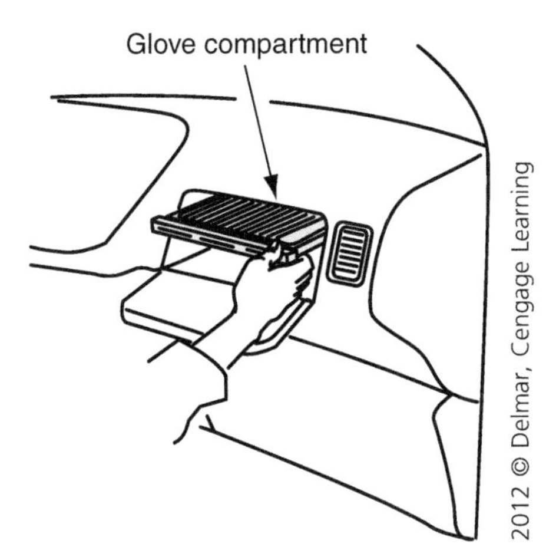

8. Referring to the figure above, what would be the HVAC malfunction if the component became restricted?

   A. AC not cold enough
   B. Heater too warm
   C. Heater not warm enough
   D. Low airflow from the ducts

   Answer A is incorrect. Problems in the refrigeration system or a misadjusted blend door could cause this problem.

   Answer B is incorrect. A misadjusted blend door could cause the heater to be warmer than selected.

   Answer C is incorrect. A faulty heater control valve or a restricted heater core could cause a heater performance problem.

   Answer D is correct. The cabin air filter cleans the duct air as it passes by it. If it gets restricted, then the airflow at the ducts will be reduced.

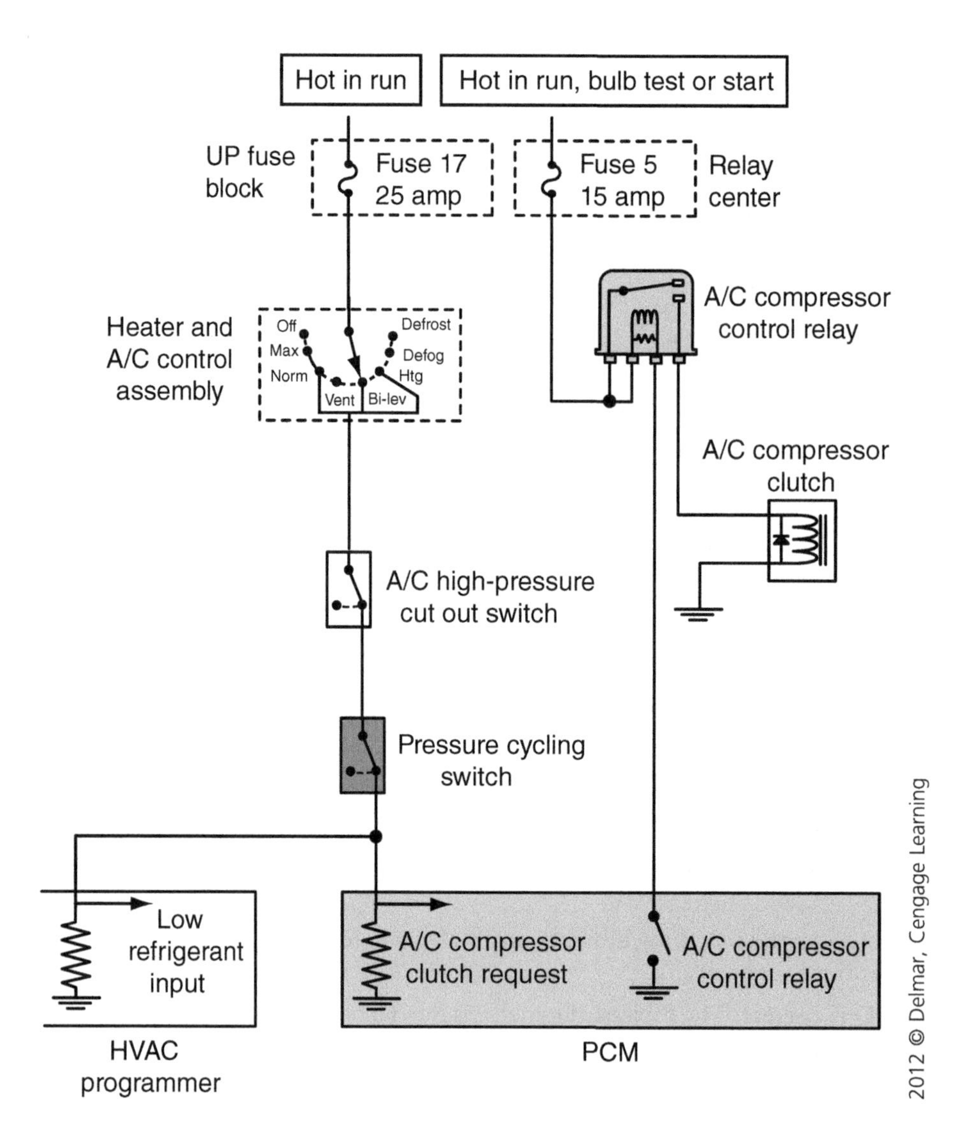

TASK C.1.3

9. Referring to the figure above, the A/C compressor clutch will not engage. Technician A says that the A/C clutch relay could be bypassed with a fused jumper to test the load side of the circuit. Technician B says that fuse 17 sends power to the A/C compressor control relay. Who is correct?

A. A only
B. B only
C. Both A and B
D. Neither A nor B.

Answer A is correct. Only Technician A is correct. Bypassing a relay with a fused jumper is a common diagnostic routine. This helps the diagnostic process by testing the load side of the system. If the compressor engages during this test, then the load side of the circuit is proven to work correctly.

Answer B is incorrect. Fuse 17 supplies power to the heater and A/C control assembly. Fuse 5 supplies power to the A/C compressor control relay.

Answer C is incorrect. Only Technician A is correct.

Answer D is incorrect. Technician A is correct.

10. A hissing noise is heard in the engine compartment for approximately one minute after the engine is shut off. Technician A says that the A/C system equalization makes an audible noise when the engine is shut off. Technician B says the heater core makes an audible noise when the engine is shut off. Who is correct?

TASK A.3

A. A only
B. B only
C. Both A and B
D. Neither A nor B

Answer A is correct. Only Technician A is correct. The pressures on both sides of the refrigeration system equalize after shutting the engine off. This process can sometimes be heard in the engine compartment.

Answer B is incorrect. The heater core does not typically make any noise when the engine is shut off.

Answer C is incorrect. Only Technician A is correct.

Answer D is incorrect. Technician A is correct.

11. The vacuum check valve was missing from the HVAC vacuum circuit but the remaining hoses were still connected. Technician A says that the modes would be stuck in the heat mode and not move in any other positions. Technician B says that the mode actuators would work normally until the vehicle drove up steep hills. Who is correct?

TASK D.2.4

A. A only
B. B only
C. Both A and B
D. Neither A nor B

Answer A is incorrect. A missing vacuum check valve would not cause the modes to be stuck in heat mode.

Answer B is correct. Only Technician B is correct. The HVAC system would still operate normally until a low vacuum situation such as driving up steep hills or other activity that requires high throttle angles.

Answer C is incorrect. Only Technician B is correct.

Answer D is incorrect. Technician B is correct.

12. A performance test of the A/C system is being performed on a late-model vehicle. Technician A says that a thermometer should be installed into the center duct during the test. Technician B says that the A/C refrigerant should be recovered during the test. Who is correct?

TASK A.1

A. A only
B. B only
C. Both A and B
D. Neither A nor B

Answer A is correct. Only Technician A is correct. One part of an A/C performance test is checking the duct temperature at the center duct with the A/C on maximum and the blower on high.

Answer B is incorrect. It is not necessary to recover the A/C refrigerant during a performance test.

Answer C is incorrect. Only Technician A is correct.

Answer D is incorrect. Technician A is correct.

TASK A.2

13. Technician A says a restricted orifice tube will cause elevated pressures on the low and high sides of the A/C system. Technician B says that faulty reed valves in the compressor can cause elevated pressures on the low and high side of the A/C system. Who is correct?

A. A only
B. B only
C. Both A and B
D. Neither A nor B

Answer A is incorrect. A restricted orifice tube would cause low pressures on both sides of the A/C refrigerant system.

Answer B is incorrect. Faulty compressor reed valves will cause the low-side pressures to be higher than normal and the high-side pressures will be lower than normal.

Answer C is incorrect. Neither Technician is correct.

Answer D is correct. Neither Technician is correct. A restricted orifice tube will not cause the pressures to elevate. Faulty compressor reed valves will not cause elevated pressures on both sides of the system.

TASK C.1.5

14. Technician A says that an overheating engine can cause the A/C to shut down and blow warm air. Technician B says that running the vehicle at wide-open throttle (WOT) for several miles can cause the A/C to lose performance. Who is correct?

A. A only
B. B only
C. Both A and B
D. Neither A nor B

Answer A is incorrect. Technician B is also correct.

Answer B is incorrect. Technician A is also correct.

Answer C is correct. Both Technicians are correct. The ECM has a strategy to de-energize the A/C compressor if the engine gets near an overheat temperature. It is also common to de-energize the A/C compressor during high throttle angle events.

Answer D is incorrect. Both Technicians are correct.

TASK A.4

15. A vehicle with an R134a A/C system is being leak tested. Which of these methods is the LEAST LIKELY method to find a refrigerant leak?

A. Visual inspection
B. Electronic detector
C. Nitrogen injection
D. Propane injection

Answer A is incorrect. Performing a good visual inspection reveals some A/C leaks. The technician should look for damaged components as well as oily residue on lines and hoses.

Answer B is incorrect. An electronic leak detector is a very helpful way to find refrigerant leaks.

Answer C is incorrect. Injecting about 120 psi of nitrogen into an A/C refrigerant system will sometimes expose a leak in the system.

Answer D is correct. Propane injection is not a normal method to detect leaks in the A/C system.

16. An A/C system that has been worked on at another shop is being diagnosed. It is suspected that a refrigerant containing "stop leak" has been added to the system. Technician A says that the refrigerant identifier will detect this substance. Technician B says that this substance can damage A/C recovery/recharging machines. Who is correct?

TASK A.7

A. A only
B. B only
C. Both A and B
D. Neither A nor B

Answer A is incorrect. Electronic refrigerant identifiers are not usually capable of detecting the sealer in an A/C system.

Answer B is correct. Only Technician B is correct. Refrigerant sealer can cause extensive damage to A/C recovery/recharge equipment.

Answer C is incorrect. Only Technician B is correct.

Answer D is incorrect. Technician B is correct.

17. A pressure switch needs to be replaced on a late-model vehicle. Technician A says that the A/C refrigerant does not have to be recovered if the switch has a Schrader valve under it. Technician B says that the refrigerant does need to be recovered if the switch is retained by a snap ring. Who is correct?

TASK B.1.2

A. A only
B. B only
C. Both A and B
D. Neither A nor B

Answer A is incorrect. Technician B is also correct.

Answer B is incorrect. Technician A is also correct.

Answer C is correct. Both Technicians are correct. A pressure switch can be serviced without recovering the refrigerant if there is a Schrader valve located under it. The Schrader prevents the loss of refrigerant when the switch is removed. On the other hand, if a switch is retained by a snap ring, then the refrigerant typically will need to be recovered since there is not a valve below the switch.

Answer D is incorrect. Both Technicians are correct.

18. A vehicle with automatic temperature control (ATC) has a "no bus" message on the control head. Technician A says that the A/C fuse is likely blown and is causing this message. Technician B says that the HVAC system may move to defrost position with heated air from the duct. Who is correct?

TASK C.3.11

A. A only
B. B only
C. Both A and B
D. Neither A nor B

Answer A is incorrect. A blown A/C fuse would not cause the "no bus" message on the control head. The data bus is wired throughout the whole vehicle and affects many different electronic systems.

Answer B is correct. Only Technician B is correct. Electronic HVAC systems will sometimes default to the heated defrost position due to safety precautions of having a clear windshield.

Answer C is incorrect. Only Technician B is correct.

Answer D is incorrect. Technician B is correct.

TASK B.1.4

19. Any of these items need to be checked and inspected when replacing an A/C compressor EXCEPT:

    A. The piston ring end clearance
    B. The clutch air gap
    C. The number of pulley grooves
    D. The location of the mounting holes

    Answer A is correct. Checking the end clearance would require the compressor to be completely disassembled.

    Answer B is incorrect. The clutch air gap needs to be checked with a feeler gauge during a compressor replacement procedure.

    Answer C is incorrect. The pulley groove count should be the same on the replacement compressor as the old compressor.

    Answer D is incorrect. The compressor mounting locations and brackets should match the old compressor.

TASK C.2.1

20. Which tool should be used to test for problems in a vacuum-operated HVAC system?

    A. Vacuum pump
    B. Pressure gauge
    C. Pressure pump
    D. Vacuum gauge

    Answer A is correct. A vacuum pump is necessary to test for many different problems in the vacuum-operated components.

    Answer B is incorrect. A pressure gauge would have no value in testing the vacuum components of the HVAC system.

    Answer C is incorrect. A pressure pump would have no value in testing the vacuum components of the HVAC system.

    Answer D is incorrect. While a vacuum gauge could check vacuum supply, it would not be suitable for testing vacuum servo motors. However, the vacuum pump could be used to do both.

TASK B.1.6

21. A knocking noise is heard in the compressor area that is audible when the compressor is engaged but it stops when the compressor turns off. Technician A says that loose compressor mounting bolts could be the cause. Technician B says that a discharge line rubbing a compressor mounting bracket could be the cause. Who is correct?

    A. A only
    B. B only
    C. Both A and B
    D. Neither A nor B

    Answer A is incorrect. Technician B is also correct.

    Answer B is incorrect. Technician A is also correct.

    Answer C is correct. Both Technicians are correct. Loose compressor mounting bolts as well as a discharge line rubbing a bracket could both cause a knocking when the compressor is engaged.

    Answer D is incorrect. Both Technicians are correct.

22. The receiver/drier is being replaced on a late-model vehicle. Technician A says that the drier should be installed last to prevent it being exposed to moisture for a prolonged time. Technician B says that some refrigerant oil should be added to the drier prior to installation. Who is correct?

TASK B.2.4

A. A only
B. B only
C. Both A and B
D. Neither A nor B

Answer A is incorrect. Technician B is also correct.

Answer B is incorrect. Technician A is also correct.

Answer C is correct. Both Technicians are correct. The drier device should always be the last A/C component to be installed during a repair since it will immediately begin absorbing moisture as soon as it is unpackaged. It is also advisable to add some refrigerant oil to the drier prior to installation. Follow manufacturer's recommendations on the quantity to add.

Answer D is incorrect. Both Technicians are correct.

23. The windshield fogs up when the defroster is turned on and the cab is filled with a sweet smell. Which of these is the most likely cause?

TASK A.21

A. Blown head gasket
B. Leaking heater control valve
C. Leaking evaporator core
D. Leaking heater core

Answer A is incorrect. A blown head gasket on the engine would not cause the fogging problem or the sweet smell in the cab.

Answer B is incorrect. A leaking heater control valve would not cause any noticeable problems inside the cab.

Answer C is incorrect. A leaking evaporator core would not produce a sweet smell in the cab.

Answer D is correct. A leaking heater core can cause windshield fogging and can sometimes produce a sweet smell from the coolant.

24. Technician A says that electronic mode actuators can be recalibrated by using a scan tool. Technician B says that electronic blend actuators can be recalibrated by using a scan tool. Who is correct?

TASK C.3.10

A. A only
B. B only
C. Both A and B
D. Neither A nor B

Answer A is incorrect. Technician B is also correct.

Answer B is incorrect. Technician A is also correct.

Answer C is correct. Both Technicians are correct. Both electronic mode and blend actuators can be recalibrated using a scan tool. This process is necessary after replacing these actuators or if they have lost the calibration.

Answer D is incorrect. Both Technicians are correct.

TASK A.22

25. A vehicle's cooling system is being inspected during a 30,000-mile service. Technician A says that the radiator cap should be pressure tested to check for correct operation. Technician B says that the coolant freeze protection should be checked with a voltmeter. Who is correct?

    A. A only
    B. B only
    C. Both A and B
    D. Neither A nor B

Answer A is correct. Only Technician A is correct. The radiator cap should be tested with a pressure pump to see if it will hold the correct pressure.

Answer B is incorrect. Coolant freeze points can be checked with a hydrometer, a refractometer, or test strips. A voltmeter is used to test for electrolysis and for galvanic voltage.

Answer C is incorrect. Only Technician A is correct.

Answer D is incorrect. Technician A is correct.

TASK C.1.2

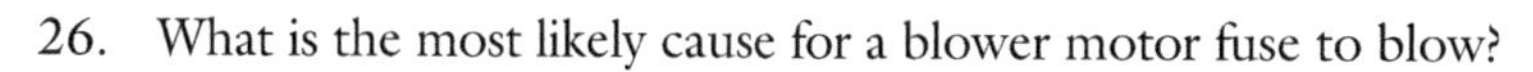

26. What is the most likely cause for a blower motor fuse to blow?

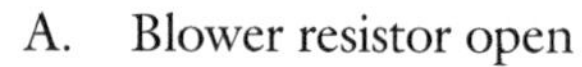

    A. Blower resistor open
    B. Faulty blower motor ground circuit
    C. Blower motor seized
    D. Burned blower relay contact

Answer A is incorrect. An open circuit causes the current flow to stop, which would never cause a fuse to blow.

Answer B is incorrect. A faulty ground would increase electrical resistance, which would reduce current flow. Reduced current flow would not cause a fuse to blow.

Answer C is correct. A seized blower motor could cause the blower fuse to blow because of the mechanical resistance to movement.

Answer D is incorrect. Burned blower relay contacts would increase electrical resistance, which would reduce current flow.

TASK A.10

27. Technician A says that a restricted cabin air filter can cause reduced airflow at all of the duct outlet locations. Technician B says that the cabin air filter should be inspected during routine preventative maintenance activities on a vehicle. Who is correct?

    A. A only
    B. B only
    C. Both A and B
    D. Neither A nor B

Answer A is incorrect. Technician B is also correct.

Answer B is incorrect. Technician A is also correct.

Answer C is correct. Both Technicians are correct. A restricted cabin air filter will reduce the airflow at all of the duct outlet locations, and this filter should be inspected during routine maintenance activities.

Answer D is incorrect. Both Technicians are correct.

28. All of the vacuum controls are inoperative with the engine running at idle speed. Technician A says the problem could be caused by a disconnected vacuum supply hose. Technician B says the manifold vacuum fitting may be blocked. Who is correct?

TASK C.2.1, C.2.2, C.2.4

A. A only
B. B only
C. Both A and B
D. Neither A nor B

Answer A is incorrect. Technician B is also correct.

Answer B is incorrect. Technician A is also correct.

Answer C is correct. Both Technicians are correct. A disconnected vacuum hose or a blocked vacuum fitting could cause all of the vacuum controls to be inoperative.

Answer D is incorrect. Both Technicians are correct.

29. What is the LEAST LIKELY cause for a heater failing to produce adequate hot air?

TASK A.20, A.28

A. Mode door out of calibration
B. Heater core with partially blocked passages
C. Heater control valve misadjusted
D. Blend air door misadjusted

Answer A is correct. A problem with a mode door will not cause temperature problems. This problem would cause an air location problem in the HVAC system.

Answer B is incorrect. A partially blocked heater core can cause reduced air temperature in a heater system. The heater core needs a good supply of hot coolant in order to provide adequate hot air for the heater.

Answer C is incorrect. A misadjusted heater control valve can cause reduced air temperature in a heater system. If the heater control valve does not fully open, then the heater core will not have enough hot coolant to do the job.

Answer D is incorrect. A misadjusted blend door can cause reduced air temperature in a heater system. The blend door directs air through or around the heater core. If this door does not fully close to direct all of the air through the heater core, then the air discharge temperature will be reduced.

30. A vehicle with automatic temperature control (ATC) has a problem with the A/C system. The set temperature and the actual temperature in the cab is 20 degrees different after 15 minutes of operation. Technician A says the cabin air temperature sensor may be defective. Technician B says the temperature blend door may be sticking. Who is correct?

TASK C.3.1

A. A only
B. B only
C. Both A and B
D. Neither A nor B

Answer A is incorrect. Technician B is also correct.

Answer B is incorrect. Technician A is also correct.

Answer C is correct. Both Technicians are correct. A defective cabin air temperature sensor or a sticking blend door could cause the actual temperature to not match the set temperature on an ATC system.

Answer D is incorrect. Both Technicians are correct.

TASK C.3.3, C.3.6

31. A vehicle has the temperature selector set to the cool position but the air from the vents is warm. The most likely problem is:

    A. The engine cooling fan is inoperative.
    B. The engine coolant level is low.
    C. The heater core's coolant passages are restricted.
    D. An air blend door is stuck.

    Answer A is incorrect. A problem with the engine coolant fan could cause engine temperature problems, but not HVAC air delivery problems.

    Answer B is incorrect. Low engine coolant level would cause lack of heat.

    Answer C is incorrect. A restricted heater core would cause lack of heat.

    Answer D is correct. The blend air door controls the temperature of the air from the vents by routing air either through or around the heater core.

TASK A.17

32. Technician A says that stored refrigerant should be kept warm by providing a heat source near the storage area. Technician B says that recovered refrigerant should be kept in a DOT 39 cylinder. Who is correct?

    A. A only
    B. B only
    C. Both A and B
    D. Neither A nor B

    Answer A is incorrect. Stored refrigerant should never be stored near a heat source due to its pressure/temperature characteristic. The pressure increases as the temperature rises.

    Answer B is incorrect. Recovered refrigerant should never be kept in a DOT39 cylinder. Always use a DOT4BW cylinder to store recovered refrigerant.

    Answer C is incorrect. Neither Technician is correct.

    Answer D is correct. Neither Technician is correct.

TASK A.5

33. A refrigerant identifier is connected to an A/C system and gives the reading of 95 percent R-134a and 5 percent R-12. Technician A says that this system can be safely recovered into the R-134a recovery machine. Technician B says this system has likely been retrofitted using nonstandard procedures. Who is correct?

    A. A only
    B. B only
    C. Both A and B
    D. Neither A nor B

    Answer A is incorrect. This system has a blend of two different refrigerants, which should never be recovered into the main shop A/C equipment.

    Answer B is correct. Only Technician B is correct. This system has R-134a and R-12, which makes it likely that a nonstandard retrofit has been performed.

    Answer C is incorrect. Only Technician B is correct.

    Answer D is incorrect. Technician B is correct.

TASK A.15

34. Technician A says that all A/C repair shops are required to use SAE-approved recovery equipment. Technician B says that all individuals who service A/C systems must be certified by a recognized body on how to properly handle refrigerants. Who is correct?

    A. A only
    B. B only
    C. Both A and B
    D. Neither A nor B

    Answer A is incorrect. Technician B is also correct.

    Answer B is incorrect. Technician A is also correct.

    Answer C is correct. Both Technicians are correct. All A/C repair shops do have to use SAE-approved A/C equipment. The service technicians must also pass certification tests distributed by recognized bodies showing that they can handle A/C refrigerants properly.

    Answer D is incorrect. Both Technicians are correct.

TASK A.3

35. Technician A says that the suction line should be hot to the touch while the A/C system is operating. Technician B says that the liquid line should be hot to the touch while the system is operating. Who is correct?

    A. A only
    B. B only
    C. Both A and B
    D. Neither A nor B

    Answer A is incorrect. The suction line should be cold to the touch when the A/C system is operating.

    Answer B is correct. Only Technician B is correct. The liquid line is hot to the touch while the A/C system is operating. The liquid line contains high-pressure liquid refrigerant that has just passed through the condenser.

    Answer C is incorrect. Only Technician B is correct.

    Answer D is incorrect. Technician B is correct.

TASK B.1.2

36. The purpose of the A/C pressure sensor is to:

    A. Open and close as refrigerant pressure changes in the low side of the system
    B. Open and close as refrigerant pressure changes in the high side of the system
    C. Send a variable signal to a processor as evaporator temperature changes
    D. Send a variable signal to a processor as high-side pressure changes

    Answer A is incorrect. The pressure cycling switch opens and closes as refrigerant pressure changes in the low side of the system.

    Answer B is incorrect. The high-pressure cutoff switch opens and closes as refrigerant pressure changes in the high side of the system.

    Answer C is incorrect. The fin temperature sensor sends a variable signal to a processor as the evaporator temperature changes.

    Answer D is correct. The A/C pressure sensor sends a variable signal to a processor as high-side pressure changes. This is a more accurate method of monitoring the high-side pressure.

TASK A.5

37. The A/C system refrigerant charge amount needs to be determined on a late-model vehicle. Technician A says that the emissions label contains the refrigerant charge amount. Technician B says that the RPO decal contains the refrigerant charge amount. Who is correct?

    A. A only
    B. B only
    C. Both A and B
    D. Neither A nor B

Answer A is incorrect. The emissions label does not contain information about the A/C system.

Answer B is incorrect. The RPO decal does not contain information about the A/C refrigerant charge amount.

Answer C is incorrect. Neither Technician is correct.

Answer D is correct. Neither Technician is correct. A refrigerant decal is located under the hood area, which gives the refrigerant capacity. This information can also be found in a technical database.

TASK A.6

38. What happens to the A/C system during the evacuation process?

    A. The refrigerant is removed from the A/C system.
    B. The A/C system is pulled into a deep vacuum to remove any moisture.
    C. The refrigerant is filtered and cleaned.
    D. The refrigerant oil is removed from the A/C system.

Answer A is incorrect. The recovery process removes the refrigerant from the system.

Answer B is correct. Evacuating the A/C system does pull the system pressure down into vacuum, which allows any moisture to boil off.

Answer C is incorrect. The filtering of the refrigerant happens when recovering the refrigerant with a multi-purpose machine.

Answer D is incorrect. The only way that all of the refrigerant oil is removed from the system is by disassembling the system and draining the oil or by flushing some of the components with a solvent-based liquid.

TASK B.1.2

39. Any of these components are A/C pressure devices EXCEPT:

    A. Thermal limiter
    B. AC pressure transducer
    C. Dual pressure switch
    D. Cycling switch

Answer A is correct. The thermal limiter is not a pressure device. The thermal limiter is a temperature switch that opens to prevent the compressor from operating under high temperature extremes.

Answer B is incorrect. The A/C pressure transducer is a variable sensor that changes its signal voltage as the A/C pressure changes.

Answer C is incorrect. The dual pressure switch is located in the high side of the system and it prevents A/C compressor operation when the pressure is very low or very high.

Answer D is incorrect. The cycling switch is located in the low side of the system, and it operates by opening a set of contacts when the A/C pressure drops below a set level.

40. Any of these steps should be followed when working around airbag systems EXCEPT:

TASK A.11

    A. Store the airbag components face down on the bench when not in use.
    B. Disconnect the negative battery cable prior to beginning work.
    C. Use caution when reconnecting airbag components.
    D. Walk while holding the airbag components away from the face.

Answer A is correct. Airbag components should never be stored face down due to the possibility of it being deployed and projected up into the air.

Answer B is incorrect. It is advisable to disconnect the negative battery cable prior to beginning work in order to remove power from the system and reduce the likelihood of deploying the bag while performing repair work.

Answer C is incorrect. It is wise to use caution when connecting airbag components.

Answer D is incorrect. It is a good idea to walk while holding the airbag components away from the face for safety.

41. A vehicle A/C system is being recharged manually with a manifold set and a 30-pound cylinder of refrigerant. Technician A says that the high-side manifold valve should be open when charging with the system turned on. Technician B says that the low-side manifold valve should be closed when charging with the system turned on. Who is correct?

TASK A.8

    A. A only
    B. B only
    C. Both A and B
    D. Neither A nor B

Answer A is incorrect. It would be dangerous to charge through the high-side while running the A/C system. High pressure would have a path to pass into the charging tank and tools, which could cause the charging tank to explode.

Answer B is incorrect. The low-side valve should be open when charging the A/C system with the system turned on.

Answer C is incorrect. Neither Technician is correct.

Answer D is correct. Neither Technician is correct. The low-side valve should be open and the high-side valve should be closed when manually recharging an A/C system with the system turned on.

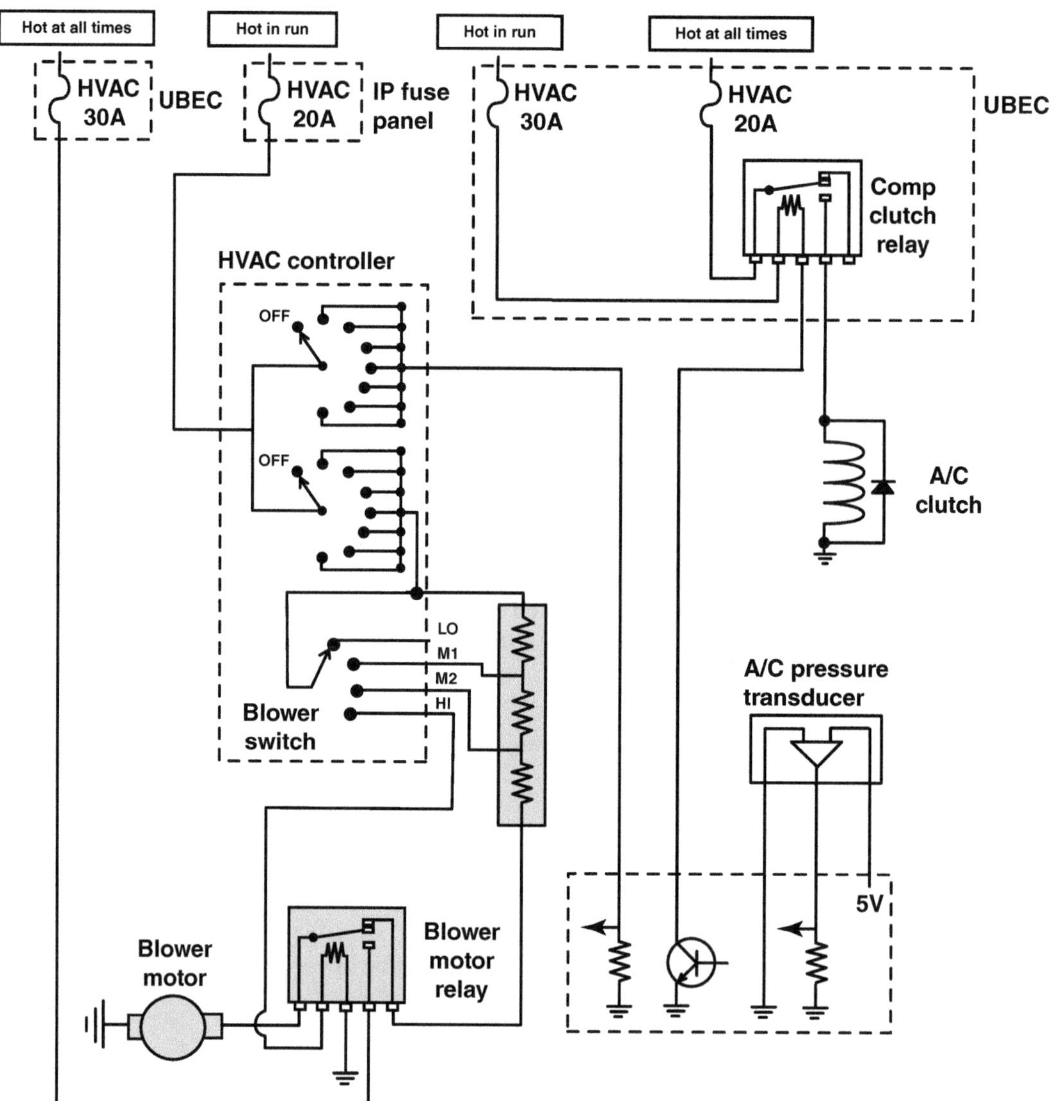

TASK C.1.2

42. Referring to the figure above, the blower will not work at any time. Technician A says that the 20 amp HVAC fuse could be blown. Technician B says that the blower motor ground could be faulty. Who is correct?

A. A only
B. B only
C. Both A and B
D. Neither A nor B

Answer A is incorrect. Technician B is also correct.

Answer B is incorrect. Technician A is also correct.

Answer C is correct. Both Technicians are correct. A blown 20 amp HVAC fuse or a bad blower motor ground could cause the blower to be totally inoperative.

Answer D is incorrect. Both Technicians are correct.

43. A serpentine belt is being replaced and the tensioner will not snap back after being released. Technician A says that the tensioner spring could be broken and the tensioner will need to be replaced. Technician B says that the idler pulley is jammed and will need to be replaced. Who is correct?

TASK B.1.3

A. A only
B. B only
C. Both A and B
D. Neither A nor B

Answer A is correct. Only Technician A is correct. The tensioner will need to be replaced if it does not snap back after it is released.

Answer B is incorrect. The idler pulley is a separate device from the tensioner pulley.

Answer C is incorrect. Only Technician A is correct.

Answer D is incorrect. Technician A is correct.

44. Technician A says that R-134a systems use mineral oil to lubricate the compressor. Technician B says that R-12 systems use PAG oil to lubricate the compressor. Who is correct?

TASK B.1.5

A. A only
B. B only
C. Both A and B
D. Neither A nor B

Answer A is incorrect. R-134a systems use PAG oil, which is a synthetic lubricant.

Answer B is incorrect. R-12 systems use mineral oil, which is a petroleum-based lubricant.

Answer C is incorrect. Neither Technician is correct.

Answer D is correct. Neither Technician is correct. R-134a systems use PAG oil and R-12 systems use mineral oil.

45. Technician A says that the HVAC control panel directs the signal to the mode actuators to determine where the air is discharged. Technician B says that some HVAC control panels contain logic capabilities. Who is correct?

TASK C.1.7, C.1.8

A. A only
B. B only
C. Both A and B
D. Neither A nor B

Answer A is incorrect. Technician B is also correct.

Answer B is incorrect. Technician A is also correct.

Answer C is correct. Both Technicians are correct. The HVAC control head does control the mode actuators by varying the signal to them. In addition, some HVAC control panels do contain logic capabilities.

Answer D is incorrect. Both Technicians are correct.

TASK A.20

46. A heater core is suspected to be restricted and causing the heater to not be warm enough. Technician A says that a temperature drop test should be performed on the heater inlet and outlet hoses. Technician B says that the heater system should create about 140°F air at the vents. Who is correct?

A. A only
B. B only
C. Both A and B
D. Neither A nor B

Answer A is incorrect. Technician B is also correct.

Answer B is incorrect. Technician A is also correct.

Answer C is correct. Both Technicians are correct. A temperature drop test on the heater inlet and outlet hoses is a good diagnostic strategy to determine if the heater core is restricted. The outlet should not be more than 15 degrees cooler than the inlet. A functional heater should produce 140 to 150°F temperature air.

Answer D is incorrect. Both Technicians are correct.

TASK C.3.9

47. Any of these steps should be followed when replacing an automatic temperature control (ATC) controller EXCEPT:

A. Disconnect the positive battery cable.
B. Connect a ground strap to your wrist and connect to a metallic component.
C. Use a battery maintainer while reprogramming the new controller.
D. Disconnect the negative battery cable.

Answer A is correct. The negative battery cable is the one that many manufacturers recommend removing to prevent having a wrench on the positive cable possibly touching a vehicle ground.

Answer B is incorrect. It is a good idea to wear a ground strap to prevent possible static damage to the controller.

Answer C is incorrect. It is a good idea to use a battery maintainer if it is necessary to reprogram the new controller. The new controller can be damaged if vehicle battery voltage gets low during a reprogram process.

Answer D is incorrect. It is a good idea to disconnect the negative battery cable when replacing control modules.

TASK A.3

48. Technician A says that the line exiting the condenser should be hotter than the line entering the condenser. Technician B says that the suction line should be cold to the touch when the A/C system is operating. Who is correct?

A. A only
B. B only
C. Both A and B
D. Neither A nor B

Answer A is incorrect. The line exiting the condenser should be from 20 to 50°F cooler than the inlet line. The condenser is a heat exchanger that causes the refrigerant to release heat, which causes the vapor to condense into a liquid.

Answer B is correct. Only Technician B is correct. The suction line is normally cold during A/C operation since it is a low-pressure vapor.

Answer C is incorrect. Only Technician B is correct.

Answer D is incorrect. Technician B is correct.

49. Before replacing an HVAC electronic control panel, the technician should:

TASK C.1.8

A. Remove the control cables from the vehicle.
B. Disconnect the negative battery cable.
C. Disassemble the dash panel.
D. Apply dielectric grease to the switch contacts.

Answer A is incorrect. There is no need to remove control cables from the vehicle before replacing any electronic control panel or device that could have solid-state circuitry.

Answer B is correct. The technician should disconnect the negative battery cable before replacing an electronic control panel. This will help prevent a voltage spike when handling the unit. The technician should also ground himself before touching the electronic panel to prevent static electricity from damaging it.

Answer C is incorrect. Some HVAC electronic control panels can be replaced without disassembling the dash panel.

Answer D is incorrect. Nothing should be applied to the switch contacts.

50. The compressor clutch will not disengage when the A/C control is switched to off. The most likely problem is:

TASK C.1.3

A. The A/C pressure cutoff switch is stuck open.
B. The compressor clutch coil is shorted to ground.
C. The low-pressure switch has an open wire.
D. The compressor coil feed circuit is shorted to voltage.

Answer A is incorrect. A stuck open A/C pressure cutoff switch would cause the compressor to never engage.

Answer B is incorrect. A short to ground on the power side of the circuit would likely blow a fuse. A short to ground on the negative side of the circuit would not have a negative effect.

Answer C is incorrect. An open wire at the low-pressure switch would keep the compressor from engaging.

Answer D is correct. A compressor coil circuit that is shorted to voltage could cause the compressor to run continuously.

PREPARATION EXAM 3—ANSWER KEY

1. D
2. B
3. D
4. A
5. A
6. C
7. A
8. A
9. B
10. D
11. B
12. C
13. B
14. C
15. D
16. C
17. C
18. B
19. B
20. D
21. C
22. C
23. C
24. C
25. A
26. B
27. B
28. C
29. B
30. B
31. D
32. A
33. D
34. A
35. C
36. C
37. A
38. A
39. A
40. B
41. C
42. A
43. B
44. C
45. D
46. B
47. D
48. C
49. B
50. C

PREPARATION EXAM 3—EXPLANATIONS

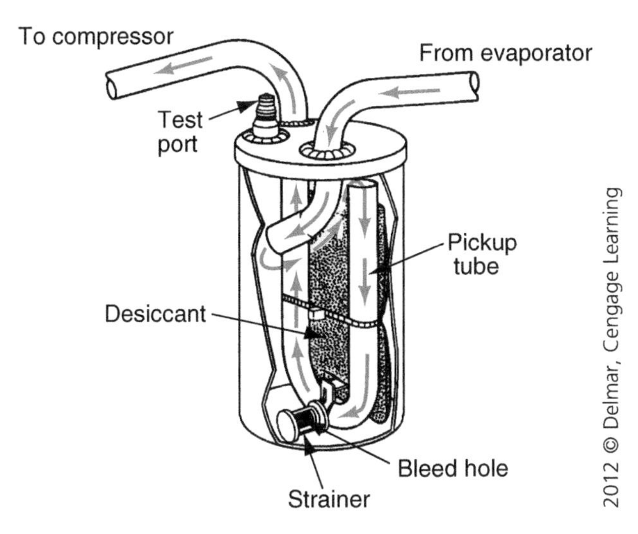

1. Which A/C system device is in the figure above?

   A. Evaporator core
   B. Receiver/drier
   C. Muffler
   D. Accumulator drier

   Answer A is incorrect. An evaporator is a heat exchanger located in the low side of the A/C system.

   Answer B is incorrect. A receiver/drier is a device used on TXV systems and is located between the condenser and the evaporator core.

   Answer C is incorrect. A muffler is a device in some discharge lines that is used to limit the sound that is created by the compressor.

   Answer D is correct. The device is an accumulator drier, which is used on orifice tube systems to absorb any moisture from the refrigerant in addition to storing vaporized refrigerant to be sent to the A/C compressor.

2. What is the LEAST LIKELY condition that could cause a blower resistor thermal element to burn up?

   A. Tight blower motor bearing
   B. Open blower motor winding
   C. Wire rubbing a metal bracket
   D. Blower motor fan blade rubbing the duct

   Answer A is incorrect. A tight bearing would cause increased current flow, which could burn the thermal element.

   Answer B is correct. An open blower motor winding would decrease the current flow and possibly disable the blower motor.

   Answer C is incorrect. A wire rubbing a metal bracket would increase current flow, which could cause the element to burn.

   Answer D is incorrect. A rubbing fan blade would increase the physical resistance, which would increase the current flow and possibly burn the element.

TASK A.1

3. An R-134a A/C system has a damaged service fitting. Technician A says the fitting needs to be mid-seated after the manifold hose adapter is installed in order to read A/C pressure. Technician B says that the Schrader valve in the fitting can be replaced without recovering the refrigerant. Who is correct?

   A. A only
   B. B only
   C. Both A and B
   D. Neither A nor B

   Answer A is incorrect. R-134a service fittings do not have to be changed in order to connect and read pressures with a manifold set. The only action that needs to take place is the manifold fitting needs to be opened to depress the Schrader valve.

   Answer B is incorrect. The refrigerant would need to be recovered before removing the Schrader valve. If the Schrader valve was removed with the refrigerant still in the system, then the whole charge would rush out of this service fitting.

   Answer C is incorrect. Neither Technician is correct.

   Answer D is correct. Neither Technician is correct. The service fitting does not have to be mid-seated to read pressures. In addition, the A/C system would need to be recovered before removing the Schrader valve.

TASK C.1.6

4. A pulse width modulated electric cooling fan motor is being diagnosed. Technician A says that the HVAC control module energizes these motors by signaling a fan control driver module. Technician B says that these fans operate at two speeds. Who is correct?

   A. A only
   B. B only
   C. Both A and B
   D. Neither A nor B

   Answer A is correct. Only Technician A is correct. The ECM typically sends a signal to a fan control driver module. The driver module delivers the current to the cooling fan motor.

   Answer B is incorrect. Pulse width modulated engine fans can have many different speeds depending on cooling load.

   Answer C is incorrect. Only Technician A is correct.

   Answer D is incorrect. Technician A is correct.

TASK A.4

5. Any of these the answer choices are a combination of instruments and methods are used to detect an A/C refrigerant leak on an R-134a system EXCEPT:

   A. Flame-type leak detector
   B. Halogen leak detector
   C. Dye and UV light
   D. Audible noise from the leak

   Answer A is correct. R-134a is not compatible with flame-type leak detectors.

   Answer B is incorrect. A halogen leak detector can be used to find R-134a leaks.

   Answer C is incorrect. Ultraviolet dye can be injected into A/C systems and will show the leak location when viewed with a UV light.

   Answer D is incorrect. Some large refrigerant leaks can be heard in a quiet area. The leak will make a hissing sound.

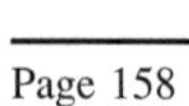

6. An A/C system has just been recovered with a multi-purpose recovery/recharging machine. Which of these statements is true of the condition of the A/C system?

TASK A.6

   A. All of the oil has been removed from the system.
   B. The system is ready for normal use.
   C. All refrigerant has been removed from the system.
   D. Humidity has been removed from the refrigerant in the system.

   Answer A is incorrect. The recovery process does not remove all of the oil from the system. In fact, very little A/C system oil is removed during the recovery.

   Answer B is incorrect. The system would not be ready for use since it will not have any refrigerant.

   Answer C is correct. The recovery process removes the refrigerant from the A/C system on the vehicle. Very little oil is removed from the A/C system during this process.

   Answer D is incorrect. The accumulator drier or receiver/drier device removes moisture from the working A/C system. The technician can remove moisture from an empty system by evacuating the system with the A/C machine or a stand-alone vacuum pump.

7. A vehicle with electronic dual-zone climate control has a problem of the passenger side only blowing hot air no matter what the passenger setting is adjusted to. Technician A says that passenger blend door actuator could be defective. Technician B says that the passenger defroster mode actuator could be defective. Who is correct?

TASK C.3.3

   A. A only
   B. B only
   C. Both A and B
   D. Neither A nor B

   Answer A is correct. Only Technician A is correct. Dual-zone climate control systems always have temperature blend doors for each side of the vehicle. A defective passenger blend door actuator is likely the cause of the problem of the passenger side only blowing hot air.

   Answer B is incorrect. A defective passenger mode actuator would affect the air location, not the temperature of the air on the passenger side.

   Answer C is incorrect. Only Technician A is correct.

   Answer D is incorrect. Technician A is correct.

8. Technician A says that automatic temperature control (ATC) control heads are typically a digital design. Technician B says that some automatic temperature control (ATC) systems use a cable-operated heat mode actuator. Who is correct?

TASK C.1.8

   A. A only
   B. B only
   C. Both A and B
   D. Neither A nor B

   Answer A is correct. Only Technician A is correct. ATC systems normally use a digital control head. This allows the driver to choose a desired temperature and the ATC system electronically achieves the set temperature.

   Answer B is incorrect. ATC systems would not normally use cable-operated doors of any type.

   Answer C is incorrect. Only Technician A is correct.

   Answer D is incorrect. Technician A is correct.

TASK A.4

9. A vehicle is being evacuated after an A/C repair but the system will not drop below 10 in. Hg after 20 minutes. Technician A says that this is normal and to vacuum for 10 more minutes. Technician B says there is likely a leak in the refrigerant system. Who is correct?

A. A only
B. B only
C. Both A and B
D. Neither A nor B

Answer A is incorrect. This is not a normal situation, and running the vacuum pump longer would be a waste of time.

Answer B is correct. Only Technician B is correct. Since the vacuum level will not drop any lower than 10 in. Hg in 20 minutes, there must be a leak somewhere in the A/C system or with the manifold and hoses.

Answer C is incorrect. Only Technician B is correct.

Answer D is incorrect. Technician B is correct.

TASK A.7

10. Which of these equipment is the most likely way to detect sealer in an A/C refrigerant system?

A. Recovery machine
B. Vacuum pump
C. Manifold set
D. Orifice bleed detection kit

Answer A is incorrect. The recovery machine removes the refrigerant from the vehicle A/C system.

Answer B is incorrect. The vacuum pump drops the system pressure into vacuum, which causes any moisture to vaporize and be removed from the A/C system.

Answer C is incorrect. The manifold set is used to measure system pressures. The manifold set is also used to add refrigerant or to remove refrigerant when used in conjunction with a recovery machine.

Answer D is correct. This is the way to test for sealer. This kit basically creates a controlled leak and allows the refrigerant to be exposed to water. If the leak rate decreases within three minutes, then sealer is likely present in the refrigerant.

TASK A.10

11. Technician A says that a restricted cabin air filter will cause reduced fuel economy. Technician B says that a cabin air filter prevents foreign debris from getting on the evaporator core. Who is correct?

A. A only
B. B only
C. Both A and B
D. Neither A nor B

Answer A is incorrect. A restricted cabin air filter would not reduce fuel economy, but it would reduce the airflow in the HVAC duct box.

Answer B is correct. Only Technician B is correct. The cabin air filter prevents debris from getting on the evaporator core. It also helps the cabin air delivery system to be much cleaner and healthier for the occupants of the vehicle.

Answer C is incorrect. Only Technician B is correct.

Answer D is incorrect. Technician B is correct.

12. During a diagnosis of an A/C system, oil residue is found around the high-pressure relief valve and the A/C refrigerant charge has been determined to be low. Technician A says the cause could be restricted air passages through the condenser. Technician B says the refrigerant system may have been overcharged. Who is correct?

TASK B.1.1

A. A only
B. B only
C. Both A and B
D. Neither A nor B

Answer A is incorrect. Technician B is also correct.

Answer B is incorrect. Technician A is also correct.

Answer C is correct. Both Technicians are correct. The high-pressure relief valve is a mechanical pressure relief valve that releases pressure when there is a system malfunction. Restricted condenser air passages or a refrigerant overcharge would both cause the high-side pressures to elevate, which could cause the high-pressure relief valve to release.

Answer D is incorrect. Both Technicians are correct.

13. The air gap on a compressor clutch is found to be twice the specified amount. Technician A says this could cause an intermittent scraping noise with the engine running and the compressor clutch disengaged. Technician B says this could cause a slipping compressor clutch when the compressor is engaged. Who is correct?

TASK B.1.4

A. A only
B. B only
C. Both A and B
D. Neither A nor B

Answer A is incorrect. An air gap that is too wide would not cause any noticeable problems when the compressor is turned off.

Answer B is correct. Only Technician B is correct. An air gap that is too wide would likely cause the compressor clutch to slip when engaged.

Answer C is incorrect. Only Technician B is correct.

Answer D is incorrect. Technician B is correct.

14. A knocking sound is heard from the area of the A/C compressor when in operation. When the compressor is shut off, the noise stops. The A/C system cools well, and there are no indications of A/C system problems. Technician A says the noise could be caused by a broken compressor mounting bracket. Technician B says that the noise could be caused by loose fasteners of the compressor mounting. Who is correct?

TASK B.1.6

A. A only
B. B only
C. Both A and B
D. Neither A nor B

Answer A is incorrect. Technician B is also correct.

Answer B is incorrect. Technician A is also correct.

Answer C is correct. Both Technicians are correct. A knocking noise that is present when the compressor is turned on could be either a broken compressor mounting bracket or loose fasteners.

Answer D is incorrect. Both Technicians are correct.

TASK B.2.2

15. A condenser is being diagnosed for an airflow problem. Technician A says a slipping fan clutch will reduce ram airflow. Technician B says reduced airflow across the condenser results in low suction pressure. Who is correct?

A. A only
B. B only
C. Both A and B
D. Neither A nor B

Answer A is incorrect. Ram airflow is present when the vehicle is moving at highway speed. A slipping fan clutch will reduce airflow at low speeds and while parked.

Answer B is incorrect. Reduced condenser airflow will cause elevated pressures on the low side and the high side.

Answer C is incorrect. Neither Technician is correct.

Answer D is correct. Neither Technician is correct. Ram airflow is present when the vehicle is moving at highway speed. A slipping fan clutch will reduce airflow at low speeds and while parked. Reduced condenser airflow will cause elevated pressures on the low side and the high side.

TASK A.3, A.2

16. A vehicle is being diagnosed for an air conditioning system that cools satisfactorily during the early morning or late evening, but does not cool during the hot part of the day. During the performance, the low-side gauge starts out reading normal, then drops into vacuum. Technician A says that ice could be forming in the expansion valve. Technician B says that the drier could be saturated with moisture. Who is correct?

A. A only
B. B only
C. Both A and B
D. Neither A nor B

Answer A is incorrect. Technician B is also correct.

Answer B is incorrect. Technician A is also correct.

Answer C is correct. Both Technicians are correct. This problem could be caused by excessive moisture in the A/C system, which could cause the expansion valve to develop ice that would restrict the flow of refrigerant. In addition, the drier could become saturated with moisture.

Answer D is incorrect. Both Technicians are correct.

TASK B.1.4

17. A vehicle is being diagnosed for the problem of the compressor clutch failing to engage with the A/C turned on. A voltage measurement shows there is 12 volt at the clutch coil feed wire. Technician A says that the clutch coil could be bad. Technician B says that the clutch coil ground could be bad. Who is correct?

A. A only
B. B only
C. Both A and B
D. Neither A nor B

Answer A is incorrect. Technician B is also correct.

Answer B is incorrect. Technician A is also correct.

Answer C is correct. Both Technicians are correct. A bad clutch coil or a bad ground could cause the compressor to not engage while receiving voltage. The technician should check the ground by performing a voltage drop test and if it is okay, then the coil would have to be the fault.

Answer D is incorrect. Both Technicians are correct.

18. What is the most likely A/C problem that will occur if the capillary tube on the TXV gets broken?

TASK B.2.5

    A. The air conditioning will always blow at full cold.
    B. The air conditioning will not blow cold enough.
    C. It will become impossible to charge the air conditioning.
    D. It will not affect the air conditioning.

Answer A is incorrect. A stuck blend door could cause the HVAC system to always blow cold air.

Answer B is correct. The capillary tube of the TXV is the feedback device. If it gets broken then the TXV will not have the capability to regulate refrigerant into the evaporator core.

Answer C is incorrect. The A/C system could still be recharged if the capillary tube gets damaged. It would just not cool well.

Answer D is incorrect. A broken capillary tube would negatively affect the air conditioning performance because it would not be capable of regulating refrigerant correctly.

19. The blower motor fuse blows each time the blower switch is turned on. Technician A says that an open low-side switch could be the cause. Technician B says that a shorted blower motor relay could be the cause. Who is correct?

TASK C.1.2, C.3.1

    A. A only
    B. B only
    C. Both A and B
    D. Neither A nor B

Answer A is incorrect. An open switch would cause current flow to stop, which would not cause a fuse to blow.

Answer B is correct. Only Technician B is correct. A shorted blower motor relay would increase current flow, which could cause a fuse to blow.

Answer C is incorrect. Only Technician B is correct.

Answer D is incorrect. Technician B is correct.

20. Technician A says that a typical A/C pressure sensor has five wires. Technician B says that a typical A/C pressure sensor operates on a 24 volt signal. Who is correct?

TASK B.1.2

    A. A only
    B. B only
    C. Both A and B
    D. Neither A nor B

Answer A is incorrect. A typical A/C pressure sensor has three wires. One wire is the 5 volt supply wire, one wire is the ground return, and the third wire is the signal wire that varies as the A/C pressure changes.

Answer B is incorrect. A/C pressure sensors operate at less than battery voltage. In many cases this voltage is at 5 volts.

Answer C is incorrect. Neither Technician is correct.

Answer D is correct. Neither Technician is correct. A typical A/C pressure sensor has three wires. One wire is the 5 volt supply wire, one wire is the ground return, and the third wire is the signal wire that varies as the A/C pressure changes. A/C pressure sensors operate at less than battery voltage.

TASK B.1.3, B.1.4

21. The A/C compressor has a squealing sound when the compressor engages. Technician A says that the belt is likely not adjusted to specifications. Technician B says that the air gap on the A/C compressor clutch plate may be too wide. Who is correct?

    A. A only
    B. B only
    C. Both A and B
    D. Neither A nor B

Answer A is incorrect. Technician B is also correct.

Answer B is incorrect. Technician A is also correct.

Answer C is correct. Both Technicians are correct. A loose belt or a wide air gap on the A/C compressor clutch plate can cause a squealing sound when the compressor engages.

Answer D is incorrect. Both Technicians are correct.

TASK A.21

22. The inside of the windshield has a foggy film. Technician A says the engine coolant level should be checked. Technician B says the heater core may be leaking. Who is correct?

    A. A only
    B. B only
    C. Both A and B
    D. Neither A nor B

Answer A is incorrect. Technician B is also correct.

Answer B is incorrect. Technician A is also correct.

Answer C is correct. Both Technicians are correct. A foggy windshield is often caused by a leaking heater core. The technician should check the coolant level to maintain proper level until the repair is made.

Answer D is incorrect. Both Technicians are correct.

TASK C.1.4

23. Which device is used as the input to cause the A/C clutch to disengage under high engine temperatures?

    A. Coolant temperature sensor
    B. Throttle position sensor
    C. Oxygen sensor
    D. Power steering switch

Answer A is correct. The coolant temperature sensor is a thermistor that is used to determine engine temperature. Some manufacturers will turn off the A/C clutch when engine temperatures elevate too high.

Answer B is incorrect. The throttle position sensor is a potentiometer that is used to provide throttle angle data to the computer.

Answer C is incorrect. The oxygen sensor is used to provide feedback to the computer on how much oxygen is in the exhaust stream after the combustion process.

Answer D is incorrect. The power steering switch is used to provide a signal to the computer when power steering pressure elevates.

24. The engine overheats when the vehicle is sitting in heavy traffic. When the vehicle is driven at highway speeds, the engine operates at the normal temperature. Technician A says a broken radiator fan shroud could be the cause. Technician B says a faulty thermostatic fan clutch could be the cause. Who is correct?

TASK A.27

A. A only
B. B only
C. Both A and B
D. Neither A nor B

Answer A is incorrect. Technician B is also correct.

Answer B is incorrect. Technician A is also correct.

Answer C is correct. Both Technicians are correct. This problem could be caused by either a broken fan shroud or a faulty thermostatic fan clutch. Both of these problems would cause overheating problems at low speeds. The radiator cools itself with forced (ram) airflow when the vehicle is moving at highway speeds.

Answer D is incorrect. Both Technicians are correct.

25. Any of these conditions could cause the cooling system to develop a voltage potential EXCEPT:

TASK A.22

A. Faulty water pump
B. A poor blower motor ground
C. Coolant acidity is too high
D. Loose negative battery cable connection at the engine block

Answer A is correct. A faulty water pump would not cause the cooling system to develop a voltage. Acidic coolant or electrical problems can cause the cooling system to develop a voltage.

Answer B is incorrect. A poor blower motor ground could cause the cooling system to act as a current path and cause a voltage to be present. This condition is called electrolysis and can cause component failure in the cooling system.

Answer C is incorrect. Highly acidic coolant can cause a galvanic action in the coolant, which can cause a voltage to be present.

Answer D is incorrect. A loose negative battery cable could cause the cooling system to act as a current path and cause a voltage to be present. This condition is called electrolysis and can cause component failure in the cooling system.

26. The blower motor operates slower than normal at all speed settings. A voltage test is performed at the blower motor connector with the switch in the high-speed switch position and 7.5 volts is measured. Technician A says that the cause could be a bad blower motor. Technician B says the problem could be a high resistance at the blower motor ground. Who is correct?

TASK C.1.2

A. A only
B. B only
C. Both A and B
D. Neither A nor B

Answer A is incorrect. The blower motor should be dropping about 13.5 volts on high with the engine running. Only 7.5 volts is being dropped on high, so there must be a voltage loss somewhere in the circuit.

Answer B is correct. Only Technician B is correct. A bad blower ground could be a point of voltage loss in this circuit. A voltage drop test of the ground circuit would need to be performed.

Answer C is incorrect. Only Technician B is correct.

Answer D is incorrect. Technician B is correct.

TASK A.1, A.2

27. During an A/C performance test on a TXV system that uses R-134a on a 78°F (25.6°C) day after 12 minutes of A/C operation, the pressure on the high-side gauge was 350 psi (2406.9 kPa) and the pressure on the low-side gauge was 50 psi (343.8 kPa). Technician A says that these readings are normal for the temperature and conditions. Technician B says that these readings could be caused by a refrigerant overcharge. Who is correct?

    A. A only
    B. B only
    C. Both A and B
    D. Neither A nor B

    Answer A is incorrect. Normal pressures on a 78°F (25.6°C) day would be about 25 to 30 psi (172.4 to 206.8 kPa) on the low side and about 150 to 175 psi (1034.2 to 1206.6 kPa) on the high side.

    Answer B is correct. Only Technician B is correct. A refrigerant overcharge could cause the elevated low- and high-side pressures seen on the gauges.

    Answer C is incorrect. Only Technician B is correct.

    Answer D is incorrect. Technician B is correct.

TASK C.3.1, C.3.5, C.3.10

28. A vehicle with an automatic temperature control (ATC) A/C system is being diagnosed. Technician A says that some systems have to be recalibrated after the vehicle battery is replaced. Technician B says that these systems use a sun load sensor to calculate radiant heat load. Who is correct?

    A. A only
    B. B only
    C. Both A and B
    D. Neither A nor B

    Answer A is incorrect. Technician B is also correct.

    Answer B is incorrect. Technician A is also correct.

    Answer C is correct. Both Technicians are correct. System calibration is sometimes needed after the vehicle battery has been unhooked. This process can be completed by using a scan tool and utilizing the bi-directional function. Some ATC systems will recalibrate by activating the internal diagnostic mode, which is typically done by pressing the control head buttons in the correct sequence. In addition, ATC systems use a sun load sensor to calculate radiant heat load.

    Answer D is incorrect. Both Technicians are correct.

TASK A.1

29. An air conditioning system equipped with an accumulator drier is a:

    A. Thermostatic expansion valve system
    B. Fixed orifice tube system
    C. Receiver/drier system
    D. Automatic temperature control system

    Answer A is incorrect. A thermostatic expansion valve system uses a receiver/drier.

    Answer B is correct. Orifice tube A/C systems use an accumulator drier that is mounted between the evaporator core and compressor. The job of the accumulator drier is to absorb any moisture that may be present in the refrigerant system in addition to storing vaporized refrigerant to be sent to the A/C compressor.

    Answer C is incorrect. The receiver/drier is a device used on TXV systems and is located between the condenser and the evaporator core.

    Answer D is incorrect. Automatic temperature control systems are electronically controlled HVAC systems that can use either a TXV or an orifice tube as the metering device.

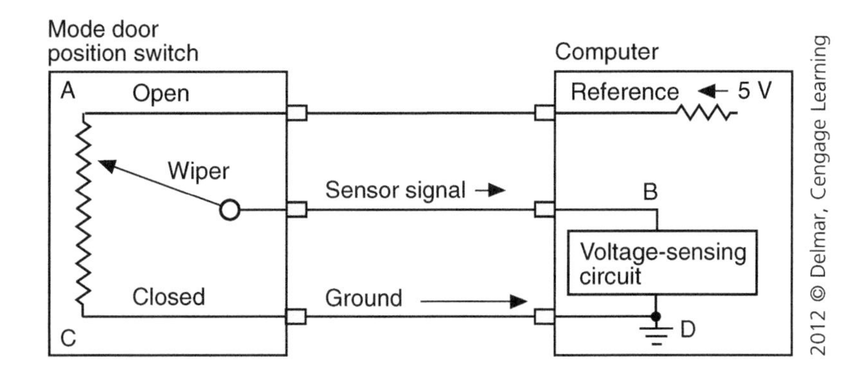

30. Referring to the figure above, the ground terminal at the computer is disconnected. Technician A says that the mode door will still operate but will set a code in the memory. Technician B says that the sensor signal will read 5 volts all of the time, no matter where the mode door is. Who is correct?

TASK C.1.7

A. A only
B. B only
C. Both A and B
D. Neither A nor B

Answer A is incorrect. If the ground is not connected, the computer has no feedback about where the door is, which would cause the system to malfunction.

Answer B is correct. Only Technician B is correct. The sensor signal would read 5 volts constantly if the ground is lost at the computer. The lost ground would create an open circuit, which would cause 5 volts to be present at all locations in the circuit.

Answer C is incorrect. Only Technician B is correct.

Answer D is incorrect. Technician B is correct.

31. Technician A says that a dual-zone climate control system uses two heater cores. Technician B says that a dual-zone climate control system uses two evaporator cores. Who is correct?

TASK C.1.7

A. A only
B. B only
C. Both A and B
D. Neither A nor B

Answer A is incorrect. A dual-zone climate control system will use two blend air doors in order to provide the driver and passenger the desired air temperature. Two heater cores are not normally used on dual-zone systems.

Answer B is incorrect. A dual-zone climate control system will use two blend air doors in order to provide the driver and passenger the desired air temperature. Two evaporator cores are not normally used on dual-zone systems.

Answer C is incorrect. Neither Technician is correct.

Answer D is correct. Neither Technician is correct.

TASK A.2

32. Technician A says that high ambient temperature levels will cause the high-side pressure to increase. Technician B says that high ambient temperature levels will cause the low-side pressure to decrease. Who is correct?

A. A only
B. B only
C. Both A and B
D. Neither A nor B

Answer A is correct. Only Technician A is correct. When the outside (ambient) temperature is elevated, the A/C system pressure will normally increase. The high-side pressure is very responsive to high ambient temperatures.

Answer B is incorrect. Low-side pressure will usually increase slightly when the ambient temperature increases.

Answer C is incorrect. Only Technician A is correct.

Answer D is incorrect. Technician A is correct.

TASK A.3

33. What is the most likely cause of water dripping out of a vehicle after running the A/C system and being parked on a hot and humid day?

A. Transmission cooler
B. Power steering
C. AC compressor
D. Duct box condensate tube

Answer A is incorrect. The transmission cooler should not leak water.

Answer B is incorrect. The power steering system should not leak water.

Answer C is incorrect. The A/C compressor should not leak anything during normal operation.

Answer D is correct. Water leaking on the ground is a normal situation after the A/C is operated on a hot and humid day. This water comes from the condensate tube, which is the tube at the bottom of the HVAC duct box. The reason this happens is the evaporator core removes heat from the duct box air and when this happens the moisture is also removed from this air. The moisture runs down to the bottom of the box and runs out on the ground.

TASK C.1.8

34. Technician A says that an electronic HVAC control head could be damaged by static electricity. Technician B says that static electricity can be eliminated if the technician never touches any metal objects. Who is correct?

A. A only
B. B only
C. Both A and B
D. Neither A nor B

Answer A is correct. Only Technician A is correct. Any electronic device can be damaged by static electricity. Technicians must take precautions to continually ground themselves while working with electronic components.

Answer B is incorrect. Static electricity can be discharged by touching metal objects. It is a good idea to touch metal objects as much as possible so static electricity will be released often.

Answer C is incorrect. Only Technician A is correct.

Answer D is incorrect. Technician A is correct.

35. Which of these is LEAST LIKELY to be performed during an A/C system recovery procedure?

TASK A.5

A. Refrigerant is removed from a vehicle.
B. Refrigerant is filtered by the recovery machine.
C. Most of the A/C system oil is removed from the vehicle.
D. Refrigerant is weighed by the recovery machine.

Answer A is incorrect. Refrigerant is removed from the vehicle during a recovery procedure.

Answer B is incorrect. The refrigerant is filtered by the recovery machine during a recovery procedure.

Answer C is correct. Very little of the A/C system oil is removed during a recovery procedure. The technician should inspect the oil bottle on the machine after recovering to see how much oil came out. Typically, less than one-half ounce is removed.

Answer D is incorrect. The refrigerant is weighed by the recovery machine during a recovery procedure. This is useful information because the technician will find out if the system was low on refrigerant.

36. Technician A says vacuum-operated HVAC duct actuators use a spring to cause them to move to a known position when no vacuum is applied. Technician B says that cable-operated HVAC duct doors can be adjusted by shortening and lengthening the Bowden cable. Who is correct?

TASK C.2.3, C.2.4

A. A only
B. B only
C. Both A and B
D. Neither A nor B

Answer A is incorrect. Technician B is also correct.

Answer B is incorrect. Technician A is also correct.

Answer C is correct. Both Technicians are correct. Vacuum actuators use a spring to cause them to move to a known position when no vacuum is applied. Cable-operated doors do have adjustment capabilities by shortening or lengthening the cable.

Answer D is incorrect. Both Technicians are correct.

37. Technician A says that cable-operated HVAC duct doors produce an audible sound when the doors hit each end of travel. Technician B says that vacuum-operated HVAC duct doors need to be calibrated after the battery has been disconnected. Who is correct?

TASK C.2.3, C.2.4

A. A only
B. B only
C. Both A and B
D. Neither A nor B

Answer A is correct. Only Technician A is correct. A definite audible sound should be heard when moving the control head to each end of travel. If the sound is not heard under the dash, then an adjustment may be necessary.

Answer B is incorrect. Vacuum-operated doors do not have to be changed after the battery has been disconnected. These systems do not use any logic; therefore, disconnecting the battery should not cause any problems with their operation.

Answer C is incorrect. Only Technician A is correct.

Answer D is incorrect. Technician A is correct.

TASK C.3.9

38. Technician A says that an electronic scan tool can be used to retrieve ATC diagnostic trouble codes from a vehicle. Technician B says that the ATC control head can be used to retrieve diagnostic trouble codes from the engine computer. Who is correct?

A. A only
B. B only
C. Both A and B
D. Neither A nor B

Answer A is correct. Only Technician A is correct. Scan tools can be used to retrieve codes from the ATC system in addition to the other computer-controlled systems on the vehicle such as the engine, transmission, and anti-lock brakes.

Answer B is incorrect. The ATC control head can often retrieve diagnostic trouble codes from the ATC system, but not the engine computer.

Answer C is incorrect. Only Technician A is correct.

Answer D is incorrect. Technician A is correct.

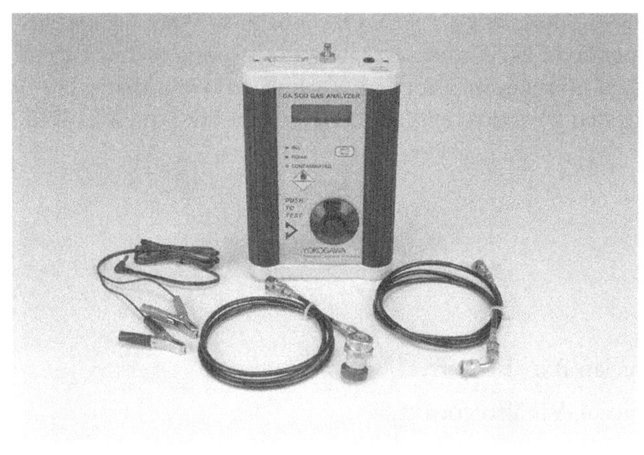

2012 © Delmar, Cengage Learning

TASK A.7

39. The tool referenced in the figure above is used:

A. To test the refrigerant for purity
B. To read A/C pressures to assist in troubleshooting for problems
C. To evacuate the A/C system to remove the moisture
D. To recover the refrigerant

Answer A is correct. The tool is a refrigerant identifier, which helps technicians test refrigerant systems for purity. It is advisable to use this important tool before connecting A/C equipment to a vehicle in order to protect against pulling mixed/impure refrigerant into the machine.

Answer B is incorrect. A tool with pressure gauges is needed to read A/C pressure.

Answer C is incorrect. A tool that incorporates a vacuum pump is needed to pull vacuum on an A/C system.

Answer D is incorrect. A recovery machine is needed to recover refrigerant.

40. A recovery machine shuts down during the recovery process. An indicator light is signaling FULL. Technician A says the car's system was overcharged and has reached the normal refrigerant level. Technician B says the recovery cylinder may be 80 percent full. Who is correct?

TASK A.5

A. A only
B. B only
C. Both A and B
D. Neither A nor B

Answer A is incorrect. The recovery machine has no way of calculating if the vehicle was overcharged.

Answer B is correct. Only Technician B is correct. Some machines will indicate that the storage tank is full when it reaches 80 percent capacity. This feature allows some expansion of the refrigerant if the temperature is increased in the shop area.

Answer C is incorrect. Only Technician B is correct.

Answer D is incorrect. Technician B is correct.

41. Frost is forming on one of the condenser tubes near the bottom of the condenser. The most likely cause of this problem would be:

TASK B.2.3

A. Restricted airflow at the condenser
B. A weak A/C compressor
C. A restricted refrigerant passage in the condenser
D. A restricted fixed orifice tube

Answer A is incorrect. A restricted airflow at the condenser would not cause frost on the surface. It would actually raise the pressure and temperature of the condenser.

Answer B is incorrect. A weak A/C compressor would cause lower than normal system pressures, but it would not cause frost at the condenser.

Answer C is correct. A restriction in the condenser could cause frost to form on the surface of the condenser. The restriction would be a point of pressure differential, which would react much like a metering device.

Answer D is incorrect. A restricted orifice tube would cause the pressures in both sides of the system to be reduced, but it would not cause frost at the condenser.

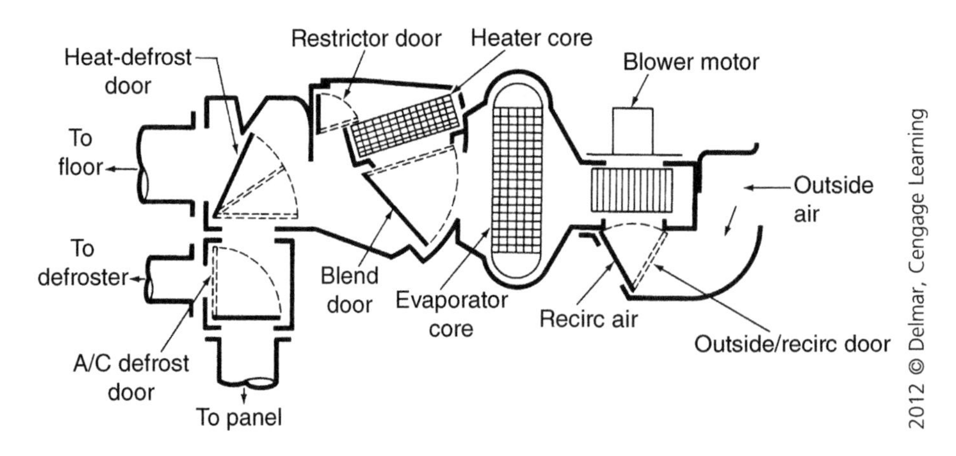

TASK C.2.5

42. Referring to the figure above, any of these statements about airflow through the duct box are correct EXCEPT:

A. The blend door routes air through or around the evaporator core.
B. The outside/recirc door position affects the origin of the air flowing through the box.
C. The A/C defrost door routes the air through to the defroster or to the panel vents.
D. The blower motor forces air through the passages and is controlled by a switch.

Answer A is correct. The blend door routes air through or around the heater core. All of the air passes the evaporator core.

Answer B is incorrect. The outside/recirc door allows air to be drawn either from outside the cab in fresh mode or from inside the cab when in recirc mode.

Answer C is incorrect. The A/C defrost door routes the air to the defroster or the panel, depending on its location.

Answer D is incorrect. The blower motor does force air through the duct passages and the speed is controlled by a switch.

TASK C.3.11

43. A technician connects a scan tool to a late-model vehicle to troubleshoot a problem in the automatic temperature control (ATC) system. The scan tool gives a fault of ambient air temperature sensor "open circuit." Technician A says that the problem could be a sensor wire rubbing a sharp body component. Technician B says that the problem could be a broken sensor wire. Who is correct?

A. A only
B. B only
C. Both A and B
D. Neither A nor B

Answer A is incorrect. A wire rubbing a part of the body would likely cause a code for a "short circuit."

Answer B is correct. Only Technician B is correct. A broken sensor wire would be a possible reason for a code for an "open circuit."

Answer C is incorrect. Only Technician B is correct.

Answer D is incorrect. Technician B is correct.

44. Technician A says that a dual pressure switch serves two functions in the A/C system. Technician B says that a dual pressure switch will open if the system loses the refrigerant charge. Who is correct?

TASK B.1.2

A. A only
B. B only
C. Both A and B
D. Neither A nor B

Answer A is incorrect. Technician B is also correct.

Answer B is incorrect. Technician A is also correct.

Answer C is correct. Both Technicians are correct. The dual pressure switch serves two functions by opening when the pressure exceeds predetermined limits in addition to opening when pressure falls below predetermined limits, such as when the refrigerant leaks out of the system.

Answer D is incorrect. Both Technicians are correct.

45. What function does the A/C clutch front drive plate perform for the A/C system?

TASK C.1.3

A. Prevents the voltage spike from damaging other components
B. Assists the clutch coil in creating magnetism
C. Limits current flow to prevent the A/C fuse from overheating
D. Connects to the driven pulley when the coil is energized

Answer A is incorrect. The clamping diode prevents the voltage spike from damaging other components.

Answer B is incorrect. The drive plate does not create magnetism. Magnetism created by the coil causes the drive plate to be pulled into the driven pulley.

Answer C is incorrect. The drive plate is a mechanical component that does not affect current flow.

Answer D is correct. The A/C clutch front drive plate is connected to the driven pulley when the coil is energized. The coil creates a magnetic field that attracts the metallic drive plate to be connected to the driven pulley.

46. A vehicle is being diagnosed for engine coolant loss. The cooling system is pressurized at 15 psi (103 kPa) for 15 minutes. There is no visible sign of coolant leaks in the engine or passenger compartments, but the pressure on the tester gauge decreases to 5 psi (34 kPa). This problem could be caused by any of the following defects EXCEPT:

TASK A.22

A. A leaking heater core
B. A leaking internal transmission cooler
C. A leaking head gasket
D. A cracked cylinder head

Answer A is incorrect. There would be coolant signs in the passenger compartment and the engine area if the heater core was leaking.

Answer B is correct. A leaking transmission cooler could cause coolant to enter the transmission fluid since the transmission cooler is in the radiator.

Answer C is incorrect. There would likely be signs of coolant on the exterior of the engine or possibly in the engine oil if the head gasket was leaking.

Answer D is incorrect. There would likely be signs of coolant on the exterior of the engine or possibly in the engine oil if the cylinder head was cracked.

TASK A.15

47. Technician A says that the service advisor needs to be certified by SAE in order to write repair orders for A/C repairs. Technician B says that the shop owner must be certified by a recognized body on how to properly handle refrigerants. Who is correct?

A. A only
B. B only
C. Both A and B
D. Neither A nor B

Answer A is incorrect. The service advisor does not have to be certified in order to handle A/C repair orders.

Answer B is incorrect. The shop owner does not have to be certified on how to handle refrigerants.

Answer C is incorrect. Neither Technician is correct.

Answer D is correct. Neither Technician is correct. The shop A/C equipment must be SAE certified and the service technicians must be certified by a recognized body on how to handle refrigerants.

TASK A.5

48. A refrigerant identifier is connected to an A/C system and gives the reading of 100 percent R-134a and 0 percent R-12. Technician A says that this system can be safely recovered into the R-134a recovery machine. Technician B says that refrigerant identifiers should be used on every vehicle prior to connecting any A/C equipment. Who is correct?

A. A only
B. B only
C. Both A and B
D. Neither A nor B

Answer A is incorrect. Technician B is also correct.

Answer B is incorrect. Technician A is also correct.

Answer C is correct. Both Technicians are correct. An R-134a system that has no other substances in it can safely be recovered into the recovery machine. It is a good idea to use the identifier on every vehicle to protect the equipment from drawing a refrigerant that is not pure.

Answer D is incorrect. Both Technicians are correct.

TASK B.2.4

49. The accumulator drier is being replaced on a late-model vehicle. Technician A says that the accumulator should be installed prior to the other A/C components. Technician B says that some refrigerant oil should be added to the accumulator prior to installation. Who is correct?

A. A only
B. B only
C. Both A and B
D. Neither A nor B

Answer A is incorrect. The accumulator drier should be installed last during the repair process in order that it will not become saturated with moisture by being exposed to the atmosphere for extended time.

Answer B is correct. Only Technician B is correct. Refrigerant oil should be added to any A/C component prior to installation in order to maintain the correct oil level and to assure that oil is present throughout the system.

Answer C is incorrect. Only Technician B is correct.

Answer D is incorrect. Technician B is correct.

50. Technician A says that R-134a systems use PAG oil to lubricate the compressor. Technician B says that R-12 systems use mineral oil to lubricate the compressor. Who is correct?

TASK B.1.5

A. A only

B. B only

C. Both A and B

D. Neither A nor B

Answer A is incorrect. Technician B is also correct.

Answer B is incorrect. Technician A is also correct.

Answer C is correct. Both Technicians are correct. PAG oil is used on R-134a systems and mineral oil is used on R-12 systems.

Answer D is incorrect. Both Technicians are correct.

PREPARATION EXAM 4—ANSWER KEY

1. A
2. D
3. C
4. A
5. A
6. D
7. A
8. C
9. A
10. B
11. A
12. C
13. C
14. A
15. B
16. B
17. D
18. C
19. D
20. C
21. C
22. B
23. C
24. A
25. D
26. D
27. C
28. C
29. D
30. B
31. C
32. C
33. D
34. A
35. B
36. B
37. C
38. A
39. C
40. A
41. C
42. A
43. C
44. C
45. C
46. B
47. C
48. B
49. C
50. C

PREPARATION EXAM 4—EXPLANATIONS

TASK A.13

1. Any of these items are typically found in a wiring diagram EXCEPT:
   A. Splice locations
   B. Circuit number of a wire
   C. Connector numbers for the circuit in question
   D. Color of the wire

Answer A is correct. Wiring diagrams do not typically describe the location of the splices that are used in the circuit.

Answer B is incorrect. Wiring diagrams will usually show the circuit numbers of the wires involved in the circuit.

Answer C is incorrect. Wiring diagrams usually show the connector numbers as well as pins identifications of the wires involved in the circuit.

Answer D is incorrect. Wiring diagrams will usually show the colors of the wires involved in the circuit.

2. The air gap on a compressor clutch is found to be 0.003 inches. The specification for this compressor is 0.020 to 0.030 inches. Technician A says this could cause an intermittent scraping noise with the engine running and the compressor clutch engaged. Technician B says this could cause a slipping compressor clutch. Who is correct?

TASK B.1.4

   A. A only
   B. B only
   C. Both A and B
   D. Neither A nor B

   Answer A is incorrect. An air gap that is too small would cause a scraping noise with the engine running and the compressor turned off.

   Answer B is incorrect. If the air gap was too wide, then the compressor clutch could slip. This air gap is too small, so it would not cause a slipping clutch.

   Answer C is incorrect. Neither Technician is correct.

   Answer D is correct. Neither Technician is correct.

3. A customer states that the heater temperature is very inconsistent. It will warm up for a few minutes and then it will cool off for a few minutes. Technician A says that the coolant level should be checked for the correct level. Technician B says that a blown head gasket could cause this problem. Who is correct?

TASK A.20

   A. A only
   B. B only
   C. Both A and B
   D. Neither A nor B

   Answer A is incorrect. Technician B is also correct.

   Answer B is incorrect. Technician A is also correct.

   Answer C is correct. Both Technicians are correct. Low coolant level or a blown head gasket could cause a heater to intermittently blow warm air. Further diagnosis would have to be performed to find the cause of this problem.

   Answer D is incorrect. Both Technicians are correct.

4. The air passages through an A/C condenser are severely restricted. Technician A says this may cause refrigerant discharge from the high-pressure relief valve. Technician B says this may cause the high-side pressure and low-side pressure to be lower than normal. Who is correct?

TASK B.2.2

   A. A only
   B. B only
   C. Both A and B
   D. Neither A nor B

   Answer A is correct. Only Technician A is correct. Blocked A/C condenser air passages will cause the high-side pressures to elevate greatly. It is possible that the pressures could reach the point that would cause the high-pressure relief valve to discharge. Normally, the high-pressure switch should open and stop the compressor prior to the system reaching the pressure that would cause the valve to discharge.

   Answer B is incorrect. The system pressures would be higher than normal if the condenser was blocked.

   Answer C is incorrect. Only Technician A is correct.

   Answer D is incorrect. Technician A is correct.

TASK A.25

5. Technician A says when the engine is running with the thermostat closed the coolant is directed through a by-pass system to ensure coolant circulation. Technician B says some systems use the upper radiator hose as a thermostat by-pass. Who is correct?

   A. A only
   B. B only
   C. Both A and B
   D. Neither A nor B

Answer A is correct. Only Technician A is correct. The coolant is still circulated through the engine while the thermostat is closed. The by-pass hose is typically the route that the coolant travels.

Answer B is incorrect. The upper radiator hose is not part of the by-pass system. Coolant travels through this hose when the thermostat opens.

Answer C is incorrect. Only Technician A is correct.

Answer D is incorrect. Technician A is correct.

TASK C.1.2

6. Technician A says a loose ground wire could cause the blower fuse to burn up. Technician B says an inoperative blower could be caused by a defective compressor clutch diode. Who is correct?

   A. A only
   B. B only
   C. Both A and B
   D. Neither A nor B

Answer A is incorrect. A loose ground wire would cause current flow to be reduced. This reduced current would not cause the fuse to blow.

Answer B is incorrect. The compressor clutch diode would not have an effect on the blower motor. The diode routes the voltage spike safely back into the coil when the coil is turned off.

Answer C is incorrect. Neither Technician is correct.

Answer D is correct. Neither Technician is correct.

TASK C.3.6

7. What is the LEAST LIKELY cause of a failed electric mode actuator on an ATC-equipped vehicle?

   A. Binding blend door
   B. Stuck defrost door
   C. Binding heater door
   D. Stuck vent door

Answer A is correct. A binding blend door would cause a problem with the blend actuator.

Answer B is incorrect. A stuck defroster door could damage the mode actuator.

Answer C is incorrect. A binding heater door could damage the mode actuator.

Answer D is incorrect. A stuck vent door could damage the mode actuator.

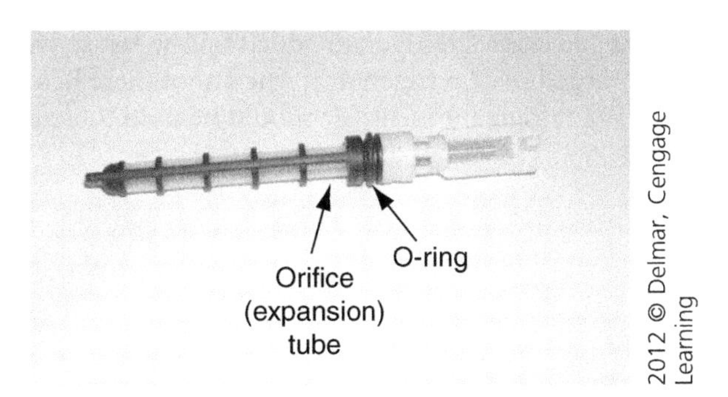

2012 © Delmar, Cengage Learning

8. In the figure above, what function does the device perform for the refrigerant system?

TASK A.1

   A. Meters refrigerant into the compressor
   B. Filters refrigerant into the condenser
   C. Meters refrigerant into the evaporator core
   D. Filters refrigerant into the compressor

   Answer A is incorrect. The suction line routes refrigerant into the compressor.

   Answer B is incorrect. The discharge line routes refrigerant into the condenser. There is not typically a filter located at this part of the system.

   Answer C is correct. The orifice tube meters high-pressure liquid refrigerant into the evaporator core. After the refrigerant passes the orifice tube, it becomes a low-pressure atomized liquid.

   Answer D is incorrect. The suction line routes refrigerant into the compressor. Some systems use a screen on the compressor inlet connection that prevents debris from entering the compressor.

9. Technician A says that a short to ground that is after the load will not cause a fuse to blow. Technician B says that an open circuit after the load will cause excessive voltage drop at the load of the circuit. Who is correct?

TASK C.1.1

   A. A only
   B. B only
   C. Both A and B
   D. Neither A nor B

   Answer A is correct. Only Technician A is correct. A short to ground that is after the load will not cause any problems if the circuit is switched on the power side. If the circuit is switched on the ground side, then the circuit will stay energized constantly due to the constant connection to ground.

   Answer B is incorrect. An open circuit will not drop voltage at any point due to the circuit being incomplete.

   Answer C is incorrect. Only Technician A is correct.

   Answer D is incorrect. Technician A is correct.

10. A vehicle is being diagnosed that had an alternative refrigerant added on the last service. Technician A says the system must be purged of all refrigerant to the atmosphere before it can be evaluated. Technician B says that a refrigerant identifier should be used to determine the type of refrigerant. Who is correct?

    A. A only
    B. B only
    C. Both A and B
    D. Neither A nor B

    Answer A is incorrect. Refrigerant should never be purged to the atmosphere.

    Answer B is correct. Only Technician B is correct. A refrigerant identifier will give feedback on what type of alternative refrigerant is being dealt with. If the vehicle does have an alternative refrigerant, the system should not be recovered in the main recovery machine. If the shop has a recovery machine that is just used for blends, then that machine can be used. If the shop does not have a machine dedicated for these types of refrigerants, then it is advisable to refuse to work on that system.

    Answer C is incorrect. Only Technician B is correct.

    Answer D is incorrect. Technician B is correct.

11. A blower motor only operates on high speed. The LEAST LIKELY cause of this condition would be?

    A. Blower motor ground is open
    B. Open wire near the blower resistor
    C. Blower switch
    D. Blower resistor

    Answer A is correct. An open blower motor ground would cause the blower motor to be inoperative on all speeds.

    Answer B is incorrect. A wire that is open near the blower resistor could cause the blower motor to only operate on high speed.

    Answer C is incorrect. A faulty blower switch could cause the blower motor to only operate on high speed.

    Answer D is incorrect. A faulty blower resistor could cause the blower to only operate on high speed.

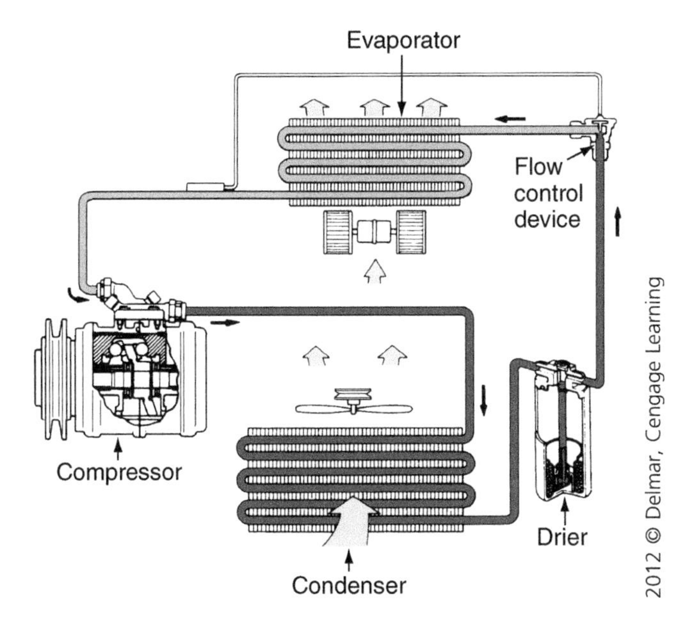

12. Referring to the figure above, what is the pressure and state of the refrigerant as it leaves the compressor?

A. High-pressure liquid
B. Low-pressure vapor
C. High-pressure vapor
D. Low-pressure liquid

Answer A is incorrect. Refrigerant would be a high-pressure liquid as it leaves the condenser.

Answer B is incorrect. Refrigerant would be a low-pressure vapor as it leaves the evaporator as well as when it enters the compressor.

Answer C is correct. The compressor increases the pressure and temperature of the refrigerant vapor so that it can condense back to a liquid when it passes through the condenser.

Answer D is incorrect. The refrigerant exists as a low-pressure liquid briefly as it leaves the expansion valve.

13. Technician A says the lines entering and exiting the receiver/drier should be the same temperature. Technician B says the lines before and after the condenser should not be the same temperature. Who is correct?

A. A only
B. B only
C. Both A and B
D. Neither A nor B

Answer A is incorrect. Technician B is also correct.

Answer B is incorrect. Technician A is also correct.

Answer C is correct. Both Technicians are correct. There should be no temperature change at the inlet or outlet of the receiver/drier. The condenser outlet line should be between 20 and 50 degrees cooler than the inlet line. The heat is given up to the air that flows through the condenser.

Answer D is incorrect. Both Technicians are correct.

TASK A.8

14. Technician A says the A/C system can be charged through the low side with the system running. Technician B says inverting the refrigerant container causes low-pressure refrigerant vapor to be charged into the system. Who is correct?

    A. A only
    B. B only
    C. Both A and B
    D. Neither A nor B

    Answer A is correct. Only Technician A is correct. It is acceptable to charge an A/C system into the low side while the system is running. Caution must be taken to make sure that the high-side manifold valves are closed during this process to make sure that high pressure does not enter the charging tank.

    Answer B is incorrect. Inverting the refrigerant container will cause low-pressure liquid to be charged into the system.

    Answer C is incorrect. Only Technician A is correct.

    Answer D is incorrect. Technician A is correct.

TASK C.3.1

15. Any of these components provide inputs to the automatic temperature control (ATC) computer EXCEPT:

    A. Sunlight sensor
    B. Engine speed sensor
    C. Cabin temperature sensor
    D. Ambient temperature sensor

    Answer A is incorrect. The sunlight sensor provides an input of sun load to the ATC computer. The system will need more cooling capacity on a bright, sunny day than on a cloudy day or at night.

    Answer B is correct. The engine speed sensor in not an input to the ATC.

    Answer C is incorrect. The cabin temperature sensor provides cabin temperature information to the ATC computer.

    Answer D is incorrect. The ambient temperature sensor provides outside air temperature information to the ATC computer.

TASK A.3

16. A vehicle makes a hissing noise each time the A/C system and engine are turned off. Technician A says that the noise is caused by a coolant leak. Technician B says that the noise is caused by equalization of system pressures. Who is correct?

    A. A only
    B. B only
    C. Both A and B
    D. Neither A nor B

    Answer A is incorrect. A coolant leak would not likely cause a hissing noise.

    Answer B is correct. Only Technician B is correct. It is normal to hear some audible noise as the A/C system equalizes each time the A/C system and the engine are turned off.

    Answer C is incorrect. Only Technician B is correct.

    Answer D is incorrect. Technician B is correct.

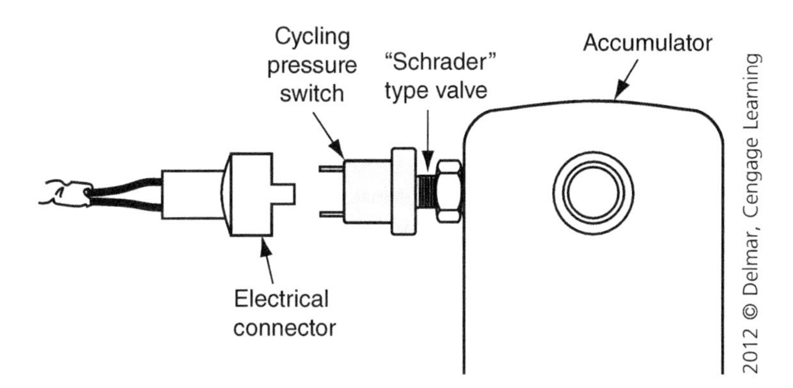

17. Referring to the figure above, Technician A says that the refrigerant will need to be recovered prior to removing the pressure cycling switch. Technician B says that the accumulator is used on TXV-type systems. Who is correct?

TASK A.1, B.1.2

A. A only
B. B only
C. Both A and B
D. Neither A nor B

Answer A is incorrect. The figure gives evidence of a Schrader valve beneath the switch, so the A/C system would not have to be recovered prior to replacing the switch.

Answer B is incorrect. Accumulators are used on systems that use an orifice tube as the metering device.

Answer C is incorrect. Neither Technician is correct.

Answer D is correct. Neither Technician is correct. The system would not have to be recovered to replace the cycling pressure switch due to having a Schrader valve beneath it. Orifice tube systems use an accumulator, and expansion valve systems use a receiver/drier.

18. A vehicle is being diagnosed for higher-than-normal system pressures. Technician A checks the condenser for dirt buildup in the condenser's fins. Technician B tests the engine for indications of an overheating condition. Who is correct?

TASK A.2

A. A only
B. B only
C. Both A and B
D. Neither A nor B

Answer A is incorrect. Technician B is also correct.

Answer B is incorrect. Technician A is also correct.

Answer C is correct. Both Technicians are correct. A blocked condenser caused by dirt could cause elevated A/C system pressures. An engine that is running hotter than normal could also cause elevated A/C system pressures.

Answer D is incorrect. Both Technicians are correct.

TASK A.8

19. Any of these conditions must be present when charging an A/C system into the high side EXCEPT:

    A. The system must be off.
    B. The system should be under vacuum.
    C. The refrigerant container should be inverted for liquid charging.
    D. The refrigerant container should be heated with a torch.

Answer A is incorrect. The A/C system should never be charged into the high side with the A/C turned on. This would be dangerous due to possibly having the high pressure exposed to the charging tank.

Answer B is incorrect. The A/C system should be evacuated prior to recharging in order to remove the moisture from the A/C system.

Answer C is incorrect. The system could be charged by inverting the container to charge with liquid.

Answer D is correct. A major heat source should never be used around a container of refrigerant. However, some older A/C machines had a small heat device that would wrap around the charging tank to raise the pressure to assist in the charging process.

TASK B.1.3

20. Technician A says that a loose air conditioning belt typically squeals as the compressor clutch is engaged. Technician B says that a slipping serpentine belt can be caused by a defective automatic tensioner. Who is correct?

    A. A only
    B. B only
    C. Both A and B
    D. Neither A nor B

Answer A is incorrect. Technician B is also correct.

Answer B is incorrect. Technician A is also correct.

Answer C is correct. Both Technicians are correct. A loose belt could cause a squeal as the A/C clutch is engaged and a slipping serpentine belt could be caused by a bad tensioner.

Answer D is incorrect. Both Technicians are correct.

TASK B.2.10

21. Technician A says that some A/C systems control evaporator pressure by varying the displacement of the compressor. Technician B says some systems regulate the flow of refrigerant into the evaporator with a thermal expansion valve (TXV). Who is correct?

    A. A only
    B. B only
    C. Both A and B
    D. Neither A nor B

Answer A is incorrect. Technician B is also correct.

Answer B is incorrect. Technician A is also correct.

Answer C is correct. Both Technicians are correct. Some manufacturers use a variable displacement compressor to control evaporator pressure. This eliminates the need to cycle the compressor to keep the system from getting too cold. Using a TXV is a way to regulate refrigerant into the evaporator, which will help control evaporator pressure, too.

Answer D is incorrect. Both Technicians are correct.

22. During a performance test of a TXV-style refrigerant system, the high-side pressure is excessively high and there is a frosted area on the condenser near the outlet. Technician A says the cause may be a refrigerant overcharge. Technician B says there may be a restriction in the condenser. Who is correct?

TASK B.2.3

A. A only
B. B only
C. Both A and B
D. Neither A nor B

Answer A is incorrect. An overcharged A/C system would not cause the condenser to have a frosted area.

Answer B is correct. Only Technician B is correct. A restriction in the condenser could cause elevated high-side pressure in addition to a cold spot on the condenser surface.

Answer C is incorrect. Only Technician B is correct.

Answer D is incorrect. Technician B is correct.

23. Any of these statements about maintaining certified A/C equipment are correct EXCEPT:

TASK A.15

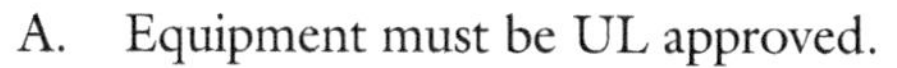

A. Equipment must be UL approved.
B. Equipment must have an approved storage container.
C. Equipment must be capable of mixing R-12 and R-134a.
D. Equipment must meet SAE standards.

Answer A is incorrect. A/C equipment must be UL approved.

Answer B is incorrect. The equipment must have an approved storage container. Machines that have replaceable storage containers must have the containers tested and recertified or be replaced every five years.

Answer C is correct. The A/C machine must never mix R-12 and R-134a.

Answer D is incorrect. The A/C equipment must meet the SAE standards when it was built. SAE continually updates the standards for A/C equipment, which technicians and shop owners need to consider when purchasing new A/C equipment. It is wise to check the current standard to make sure the new equipment meets the standard.

24. Any of these A/C components can be flushed during A/C service EXCEPT:

TASK B.2.1

A. Muffler
B. A/C suction hose
C. A/C liquid line
D. Evaporator core

Answer A is correct. A discharge line muffler is not able to be thoroughly flushed due to contaminants sticking inside the muffler assembly.

Answer B is incorrect. The suction hose can be flushed without a problem.

Answer C is incorrect. The liquid line can be flushed without a problem.

Answer D is incorrect. The evaporator core can be flushed without a problem. If a TXV valve is used, then the valve will need to be removed prior to flushing the evaporator core.

TASK B.2.3

25. A temperature drop test is being performed on a condenser. The inlet line was measured at 145°F and the outlet line temperature was measured at 120°F. Technician A says that the condenser is likely internally restricted. Technician B says that outlet temperature should be hotter than the inlet temperature. Who is correct?

A. A only
B. B only
C. Both A and B
D. Neither A nor B

Answer A is incorrect. The results of this test are normal, so there is no evidence given that the condenser is restricted.

Answer B is incorrect. The outlet temperature should be cooler than the inlet temperature.

Answer C is incorrect. Neither Technician is correct.

Answer D is correct. Neither Technician is correct. The results given from this temperature drop test are within specifications. The temperature drop should be between 20 and 50°F from the inlet to the outlet.

TASK B.1.5

26. Technician A says that a leaking A/C compressor shaft seal will not produce any visible sign of a leak. Technician B says that a compressor shaft seal can be replaced without recovering the refrigerant. Who is correct?

A. A only
B. B only
C. Both A and B
D. Neither A nor B

Answer A is incorrect. A leaking compressor shaft seal will usually produce an oily residue around the compressor front pulley and clutch assembly.

Answer B is incorrect. The refrigerant would have to be recovered prior to replacing the compressor shaft seal. Not all compressors have serviceable front seals.

Answer C is incorrect. Neither Technician is correct.

Answer D is correct. Neither Technician is correct.

TASK A.20

27. A vehicle is being diagnosed for a poor performing heater. The HVAC control is set to full heat but the temperature of the air is barely warm. Technician A says the heater core may be restricted. Technician B says the engine cooling system thermostat may be stuck open. Who is correct?

A. A only
B. B only
C. Both A and B
D. Neither A nor B

Answer A is incorrect. Technician B is also correct.

Answer B is incorrect. Technician A is also correct.

Answer C is correct. Both Technicians are correct. A restricted heater core or a stuck open thermostat could cause a heater performance problem. Further diagnosis would need to be performed before a root cause could be found.

Answer D is incorrect. Both Technicians are correct.

28. There is a noticeable noise coming from the engine compartment when the A/C is selected. Technician A says this could be caused by a loose compressor mount. Technician B says this could be caused by the discharge line touching a metal bracket. Who is correct?

TASK B.1.6

A. A only
B. B only
C. Both A and B
D. Neither A nor B

Answer A is incorrect. Technician B is also correct.

Answer B is incorrect. Technician A is also correct.

Answer C is correct. Both Technicians are correct. A loose compressor mount would cause audible noise when the A/C is selected because of compressor vibration. A discharge line that touches a metal bracket could also cause an audible noise with the A/C turned on.

Answer D is incorrect. Both Technicians are correct.

29. Any of these cooling system tests are valuable to perform during a maintenance service EXCEPT:

TASK A.22

A. pH test for acidity
B. Freeze protection
C. Coolant voltage
D. Water pump end-play

Answer A is incorrect. It is normal to check the pH level of the coolant to measure the acidity.

Answer B is incorrect. It is normal to check the freeze protection of the coolant during a coolant maintenance check.

Answer C is incorrect. Checking to see if any DC voltage is present in the coolant is a way to test for electrolysis as well as a galvanic problem possibly caused by poor coolant maintenance.

Answer D is correct. Checking the end-play of the water pump is not a common practice when performing a cooling system maintenance procedure. This test would be very labor-intensive on most vehicles.

30. Technician A says a purpose of the A/C clutch diode is to prevent alternating current from the alternator entering the clutch coil. Technician B says a purpose of the A/C clutch diode is to prevent spikes of high voltage created by the clutch coil operation from damaging delicate electronic components. Who is correct?

TASK C.1.3

A. A only
B. B only
C. Both A and B
D. Neither A nor B

Answer A is incorrect. The A/C clutch diode does not have any effect on the alternator.

Answer B is correct. Only Technician B is correct. The A/C clutch diode acts as a clamping device that routes the voltage spike back into the coil to prevent it from possibly damaging electronic components.

Answer C is incorrect. Only Technician B is correct.

Answer D is incorrect. Technician B is correct.

TASK C.1.5

31. Technician A says that some vehicles use the throttle position sensor input to disengage the A/C compressor during times of heavy load. Technician B says some vehicles have a power steering cutoff switch to disengage the A/C compressor when the power steering requires maximum effort. Who is correct?

   A. A only
   B. B only
   C. Both A and B
   D. Neither A nor B

Answer A is incorrect. Technician B is also correct.

Answer B is incorrect. Technician A is also correct.

Answer C is correct. Both Technicians are correct. Some vehicles will shut off the A/C under heavy throttle, which is monitored by the engine controller with the throttle position sensor. In addition, some vehicles utilize a power steering switch to monitor the load of the power steering system. This input is used to increase the idle and also to disengage the A/C compressor when high power steering pressure closes the switch.

Answer D is incorrect. Both Technicians are correct.

TASK C.2.4

32. Technician A says that all of the duct air typically passes through the evaporator core. Technician B says the incoming air must go through the evaporator even if heat is selected. Who is correct?

   A. A only
   B. B only
   C. Both A and B
   D. Neither A nor B

Answer A is incorrect. Technician B is also correct.

Answer B is incorrect. Technician A is also correct.

Answer C is correct. Both Technicians are correct. All of the duct air typically passes through the evaporator core and is then routed either through or around the heater core, depending on the selected temperature position.

Answer D is incorrect. Both Technicians are correct.

TASK C.1.1

33. Technician A says that an open circuit will typically cause a fuse to blow. Technician B says that a corroded connection will typically cause a fuse to blow. Who is correct?

   A. A only
   B. B only
   C. Both A and B
   D. Neither A nor B

Answer A is incorrect. Fuses blow due to high current flow but an open circuit causes current flow to stop.

Answer B is incorrect. Corroded connections add electrical resistance to a circuit, which reduces current flow.

Answer C is incorrect. Neither Technician is correct.

Answer D is correct. Neither Technician is correct. Open circuits stop current flow and corroded circuits reduce current flow. Neither of these conditions will cause fuses to blow.

34. What can a refrigerant identifier determine about a refrigerant system?

TASK A.7

   A. Air in the refrigerant system
   B. Excessive refrigerant charge
   C. Excessive oil in the refrigerant system
   D. Low refrigerant charge

   Answer A is correct. A refrigerant identifier will sample the refrigerant in an A/C system and determine the amount of air. In addition, the tool will give the percentages of R-134a, R-12, R-22, and hydrocarbon content.

   Answer B is incorrect. A recovery machine would be needed to verify the charge level in an A/C system.

   Answer C is incorrect. The identifier does not measure the amount of oil in the system. Care must be taken in order to not install excessive refrigerant oil in a system.

   Answer D is incorrect. An A/C performance test could be performed to determine if the system is low on refrigerant.

35. A blower motor switch is being inspected and the technician finds burned contacts on the switch. What would be the most likely cause of this condition?

TASK C.1.2

   A. Open blower resistor
   B. Tight blower motor bearing
   C. Loose blower motor ground
   D. High resistance across the blower relay

   Answer A is incorrect. An open blower resistor will likely cause the blower to not work on one or more of the lower speeds. This would not cause the switch contacts to get burned.

   Answer B is correct. A tight blower motor bearing will cause an increase in current flow, which could cause the contacts on the switch to get burned.

   Answer C is incorrect. A loose blower motor ground would reduce current flow, which would not contribute to burned switch contacts.

   Answer D is incorrect. A high resistance circuit at the blower relay would reduce current flow, which would not cause burned switch contacts.

36. What is the most likely tool to be used when troubleshooting a data communication network?

TASK C.3.11

   A. Continuity tester
   B. Digital multimeter
   C. 12 volt test light
   D. Short finder

   Answer A is incorrect. The only use for a continuity tester is for checking closed circuits such as short pieces of electrical wire or simple switches.

   Answer B is correct. A digital multimeter can be used on circuits that involve a logic device since it has high impedance.

   Answer C is incorrect. A 12 volt test light should not be used on a circuit that involves a logic device because it has low impedance and would cause too much current flow during its use.

   Answer D is incorrect. A short finder would be used on electrical circuits that have a problem of blowing fuses. It is not commonly used on data circuits.

TASK C.1.7

37. Technician A says that electric mode actuators use a position sensor to provide feedback to the HVAC computer about door position. Technician B says that electric blend door actuators use a position sensor to provide the HVAC computer information about the blend door position. Who is correct?

A. A only
B. B only
C. Both A and B
D. Neither A nor B

Answer A is incorrect. Technician B is also correct.

Answer B is incorrect. Technician A is also correct.

Answer C is correct. Both Technicians are correct. Both mode actuators and blend door actuators use feedback sensors to signal the HVAC controller the location of each door.

Answer D is incorrect. Both Technicians are correct.

TASK C.2.1

38. A hissing sound is heard behind the dash panel. Technician A says that the vacuum switching valve could be leaking. Technician B says that a vacuum hose could be pinched. Who is correct?

A. A only
B. B only
C. Both A and B
D. Neither A nor B

Answer A is correct. Only Technician A is correct. A vacuum switching valve that leaks could cause a hissing noise behind the dash panel.

Answer B is incorrect. A pinched vacuum hose could cause problems in the HVAC air handling system but is unlikely to cause a hissing sound.

Answer C is incorrect. Only Technician A is correct.

Answer D is incorrect. Technician A is correct.

TASK C.2.1, C.2.4

39. Any of these methods of troubleshooting vacuum problems on an HVAC duct system are correct EXCEPT:

A. Listening for hissing sounds with an ultrasonic leak detector
B. Using a vacuum pump to test vacuum actuators
C. Connecting the A/C evacuation pump to the system to monitor leaks
D. Visually inspecting hoses and actuators

Answer A is incorrect. Listening for vacuum leaks with an ultrasonic detector would be a good method of leak detection.

Answer B is incorrect. A vacuum pump is commonly used to troubleshoot the HVAC vacuum system.

Answer C is correct. It is not normal to connect the A/C evacuation pump to troubleshoot the HVAC duct system vacuum components.

Answer D is incorrect. A visual inspection of the vacuum hoses is a good way to do preliminary checks of the HVAC system vacuum.

40. Technician A says that failing to change the cabin air filter according to the maintenance schedule could result in reduced airflow from the ducts. Technician B says that the cabin air filters should be serviced every 100,000 miles. Who is correct?

TASK A.10

A. A only
B. B only
C. Both A and B
D. Neither A nor B

Answer A is correct. Only Technician A is correct. A restricted cabin air filter would result in reduced airflow from the ducts.

Answer B is incorrect. The manufacturer's recommendations should be followed concerning the cabin air filter. The intervals would likely be more frequent than every 100,000 miles.

Answer C is incorrect. Only Technician A is correct.

Answer D is incorrect. Technician A is correct.

41. Technician A says that some automatic temperature control (ATC) systems have a self-diagnostic feature in the HVAC control head. Technician B says that some automatic temperature control (ATC) systems have a calibration feature available in the HVAC control head. Who is correct?

TASK C.3.1, C.3.10

A. A only
B. B only
C. Both A and B
D. Neither A nor B

Answer A is incorrect. Technician B is also correct.

Answer B is incorrect. Technician A is also correct.

Answer C is correct. Both Technicians are correct. Some ATC control heads have a built-in diagnostic feature as well as the function to perform a calibration of the duct doors.

Answer D is incorrect. Both Technicians are correct.

42. Technician A says that some scan tools have an output test function that will energize the A/C compressor clutch. Technician B says that some scan tools have an output test that will energize the A/C pressure cycling switch. Who is correct?

TASK C.3.4

A. A only
B. B only
C. Both A and B
D. Neither A nor B

Answer A is correct. Only Technician A is correct. Some scan tools have the function in the output test mode to energize the A/C compressor clutch.

Answer B is incorrect. Testing the A/C pressure cycling switch would not happen with an output test. Some scan tools will show the condition of the switch while viewing the data list.

Answer C is incorrect. Only Technician A is correct.

Answer D is incorrect. Technician A is correct.

TASK C.1.8

43. What is the LEAST LIKELY step that would need to be followed when replacing the A/C control panel?

A. Disconnect the wiring and cables.
B. Remove the trim bezel.
C. Recover the refrigerant.
D. Disconnect the negative battery cable.

Answer A is incorrect. Disconnecting the wiring and cables is a normal step in replacing the A/C control head.

Answer B is incorrect. Removing the trim bezel is a normal step in replacing the A/C control panel.

Answer C is correct. The refrigerant would not have to be recovered to replace the A/C control panel.

Answer D is incorrect. Disconnecting the negative battery cable is a normal step in replacing some A/C control panels. This is a highly recommended step when replacing electronic A/C control panels.

TASK C.3.11

44. Technician A says that the scan tool receives data from the vehicle by communicating on the data bus network. Technician B says that if both of the data bus wires break, then the network will not communicate. Who is correct?

A. A only
B. B only
C. Both A and B
D. Neither A nor B

Answer A is incorrect. Technician B is also correct.

Answer B is incorrect. Technician A is also correct.

Answer C is correct. Both Technicians are correct. The scan tool does receive data over the data network. In addition, the data network would not be able to communicate if both data bus wires were broken.

Answer D is incorrect. Both Technicians are correct.

TASK A.5

45. Technician A says that a refrigerant identifier will determine the purity of the refrigerant. Technician B says that the refrigerant identifier should be connected to a vehicle prior to connecting the recovery station. Who is correct?

A. A only
B. B only
C. Both A and B
D. Neither A nor B

Answer A is incorrect. Technician B is also correct.

Answer B is incorrect. Technician A is also correct.

Answer C is correct. Both Technicians are correct. The refrigerant identifier tests the purity of the refrigerant and it should be connected prior to connecting the recovery machine. Some A/C recovery/recharging machines have a built-in identifier.

Answer D is incorrect. Both Technicians are correct.

46. Any of these examples are possible methods for a technician to retrieve trouble codes from a vehicle HVAC system EXCEPT:

TASK A.12

A. Connect a scan tool to the vehicle data link connector (DLC) to communicate with the HVAC system.

B. Connect a scan tool directly to the HVAC control module to communicate with the HVAC system.

C. Depress the buttons on the HVAC control head and watch the flashing indicator.

D. Depress the buttons on the HVAC control head and read the code on the electronic display.

Answer A is incorrect. Trouble codes can be retrieved by connecting a scan tool to the DLC.

Answer B is correct. The scan tool is not typically connected directly to the HVAC control module.

Answer C is incorrect. Some vehicles will display trouble codes after buttons are depressed on the HVAC control head. The codes are displayed by a flashing indicator on the control head.

Answer D is incorrect. Some vehicles will display trouble codes after buttons are depressed on the HVAC control head. The codes are displayed by an electronic display on the control head.

47. A pressure check of a tank of refrigerant that had been stored 16 hours at 70°F (21.1°C) showed 98 psi (673.95 kPa). What is the most likely factor that would cause these readings?

TASK A.18

A. Excessive refrigerant oil

B. Container is overfilled

C. Excessive air

D. Normal reading

Answer A is incorrect. Refrigerant oil content will not cause the static pressure to be elevated.

Answer B is incorrect. An overfilled container will not cause the static pressure to be elevated.

Answer C is correct. Too much air mixed with the refrigerant would cause the pressure to be higher than specifications.

Answer D is incorrect. This is not a normal reading for a 70°F day. The pressure should not exceed 76 psi.

48. Technician A says the high-pressure relief valve is a mechanical relief valve that exhausts refrigerant when the pressure exceeds 280 psi (1930.5 kPa). Technician B says that the high-pressure relief valve resets automatically after it releases pressure. Who is correct?

TASK B.2.12

A. A only

B. B only

C. Both A and B

D. Neither A nor B

Answer A is incorrect. High-pressure relief valves typically will not release pressure until a much higher point than 280 psi (1930.5 kPa). Each manufacturer will have a specification, but none would be this low due to pressures normally running this high on very hot and humid days.

Answer B is correct. Only Technician B is correct. The high-pressure relief valve is a mechanical blow-off valve that automatically resets after it releases pressure.

Answer C is incorrect. Only Technician B is correct.

Answer D is incorrect. Technician B is correct.

TASK C.2.1

49. All of the vacuum controls are inoperative with the engine running at idle speed. Technician A says the problem could be caused by a blocked vacuum supply hose. Technician B says the manifold vacuum fitting may be disconnected. Who is correct?

A. A only
B. B only
C. Both A and B
D. Neither A nor B

Answer A is incorrect. Technician B is also correct.

Answer B is incorrect. Technician A is also correct.

Answer C is correct. Both Technicians are correct. A blocked or disconnected vacuum hose could cause the vacuum-controlled HVAC components to not work.

Answer D is incorrect. Both Technicians are correct.

TASK A.5

50. Which of these processes removes the refrigerant from the A/C system?

A. Identification
B. Leak test
C. Recovery
D. Evacuation

Answer A is incorrect. A refrigerant identifier tests the purity of the refrigerant.

Answer B is incorrect. A/C systems can be leak tested by using dye and black light, electronic detector, or by visual inspection.

Answer C is correct. The recovery process removes the refrigerant from the A/C system.

Answer D is incorrect. The evacuation process puts the A/C system into vacuum in order to help remove any moisture that might be present in the system.

PREPARATION EXAM 5—ANSWER KEY

1. B
2. A
3. B
4. C
5. C
6. D
7. C
8. A
9. B
10. C
11. C
12. C
13. A
14. B
15. C
16. B
17. C
18. C
19. D
20. C
21. D
22. A
23. A
24. A
25. B
26. C
27. A
28. C
29. B
30. A
31. B
32. B
33. A
34. B
35. B
36. A
37. D
38. A
39. B
40. B
41. B
42. C
43. B
44. B
45. B
46. B
47. C
48. D
49. B
50. C

PREPARATION EXAM 5—EXPLANATIONS

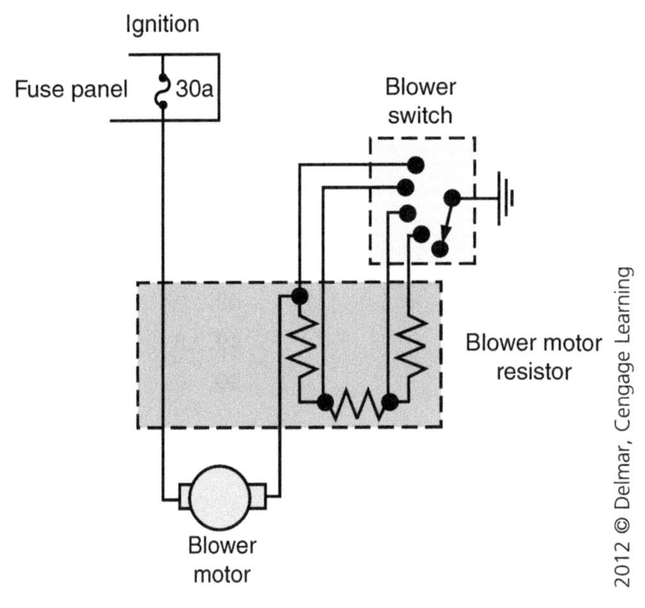

TASK C.1.2

1. How many speeds does the blower motor in the figure above have?
   A. Three
   B. Four
   C. Five
   D. Six

Answer A is incorrect. The blower motor has four speeds. See answer B for an explanation.

Answer B is correct. The blower motor has four speeds. On low speed, the path for current flow would be through the blower motor and then through all three of the blower resistors and then through the switch to ground. On medium-low speed, the path for current flow would be through the blower motor and then through two of the resistors and on through the switch to ground. On medium-high speed, the path would be through the blower motor and then through one of the resistors and on through the switch to ground. On high speed, the path of the current flow would be through the blower motor and then through the switch to ground.

Answer C is incorrect. The blower motor has four speeds See answer B for an explanation.

Answer D is incorrect. The blower motor has four speeds. See answer B for an explanation.

TASK A.3

2. Technician A says the lines before and after the receiver/drier should be the same temperature. Technician B says the lines before and after the condenser should be the same temperature. Who is correct?
   A. A only
   B. B only
   C. Both A and B
   D. Neither A nor B

Answer A is correct. Only Technician A is correct. There should not be a drop in temperature at the receiver/drier inlet and outlet lines.

Answer B is incorrect. The condenser outlet line should be about 20 to 50°F cooler than the inlet line. The condenser is a heat exchanger, which causes this temperature drop.

Answer C is incorrect. Only Technician A is correct.

Answer D is incorrect. Technician A is correct.

3. An A/C system has a vacuum reservoir, a check valve, and a vacuum-operated mode door. While operating in the A/C mode and climbing a steep hill with the throttle nearly wide open, the air discharge switches from the panel to the defroster ducts. The A/C system operates normally under all other conditions. The most likely cause of this problem would be:

TASK C.2.4

A. A leaking panel door vacuum actuator
B. A defective vacuum reservoir check valve
C. A leaking blend-air door vacuum actuator
D. A leaking intake manifold gasket

Answer A is incorrect. A leaking panel door vacuum actuator would likely fail to operate normally all the time.

Answer B is correct. A bad check valve will not trap vacuum in the HVAC system when the engine vacuum is reduced due to engine load. This will cause the mode doors to move to their spring-loaded positions.

Answer C is incorrect. Vacuum is not typically used on the blend door due to the wide variables that the blend door needs to have. Vacuum actuators will only have two or three possible positions.

Answer D is incorrect. A leaking intake manifold gasket would cause the vacuum level to be low at all times, which would cause HVAC vacuum operation to function poorly at all times.

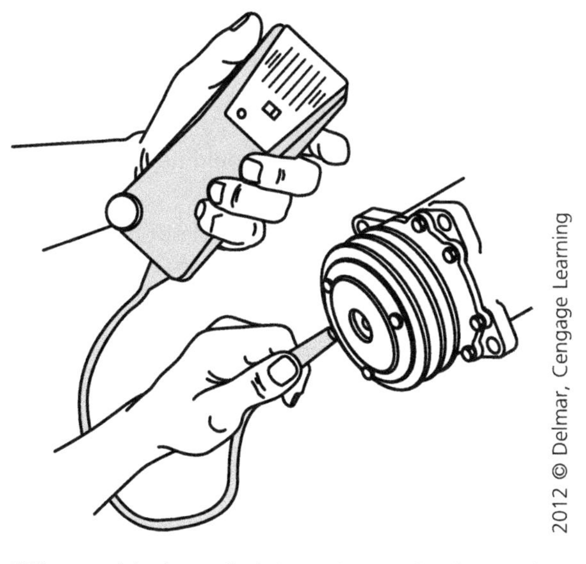

4. What tool is the technician using in the figure above?

TASK A.4

A. Refrigerant identifier
B. Belt tension gauge
C. Refrigerant leak detector
D. Electronic manifold device

Answer A is incorrect. The tool is not a refrigerant identifier. An identifier has a hose connection that connects to the service fitting on the A/C system.

Answer B is incorrect. The tool is not a belt tension gauge. A belt tension gauge is used to determine the tightness of the drive belt.

Answer C is correct. The tool in the figure is a refrigerant leak detector. This tool provides an audible sound when it senses refrigerant.

Answer D is incorrect. The tool is not an electronic manifold device. Electronic manifold sets would have the normal service hoses that analog manifolds have always used.

TASK A.1

5. Any of these faults in the A/C system could cause an elevated high-side reading EXCEPT:
   A. Refrigerant overcharge
   B. Restricted airflow to the condenser
   C. Poor airflow across the evaporator
   D. A slipping fan clutch

Answer A is incorrect. An overcharged A/C system would cause elevated high-side pressures.

Answer B is incorrect. A condenser airflow problem would cause elevated high-side pressures.

Answer C is correct. Poor airflow across the evaporator core would not cause elevated high-side pressure.

Answer D is incorrect. A slipping fan clutch could cause elevated high-side pressures when the vehicle is at low speeds or is sitting still.

TASK A.7

6. A vehicle being serviced is identified to have an alternative refrigerant in the A/C system. Technician A says the system should be evacuated before service begins. Technician B says that the refrigerant should be vented to the atmosphere and then recharged with pure refrigerant. Who is correct?
   A. A only
   B. B only
   C. Both A and B
   D. Neither A nor B

Answer A is incorrect. A system with an alternative refrigerant should not be connected to the normal A/C equipment.

Answer B is incorrect. A system with an alternative refrigerant should not be vented to the atmosphere. It is illegal to intentionally vent any type of refrigerant into the atmosphere.

Answer C is incorrect. Neither Technician is correct.

Answer D is correct. Neither Technician is correct. Either the system should be recovered using a machine that is only used for alternative refrigerants or the customer should be counseled on where the A/C can be serviced.

TASK B.2.10

7. Technician A says that some A/C systems prevent the evaporator from freezing up by varying the displacement of the compressor. Technician B says some systems regulate the flow of refrigerant into the evaporator with a variable orifice valve (VOV). Who is correct?
   A. A only
   B. B only
   C. Both A and B
   D. Neither A nor B

Answer A is incorrect. Technician B is also correct.

Answer B is incorrect. Technician A is also correct.

Answer C is correct. Both Technicians are correct. Some A/C systems use a variable displacement compressor to control the evaporator temperature. These compressors do not cycle on and off to control evaporator temperature. Some A/C systems use a VOV to control evaporator temperature. These devices operate similarly to expansion valves by having the ability to increase or decrease refrigerant flow based upon the evaporator temperature.

Answer D is incorrect. Both Technicians are correct.

8. Technician A says a vacuum pump is used during the recovery process to remove the last trace of refrigerant from the system. Technician B says any oil not recovered from the air conditioning system reduces the system's efficiency. Who is correct?

TASK A.6

   A. A only
   B. B only
   C. Both A and B
   D. Neither A nor B

   Answer A is correct. Only Technician A is correct. The A/C machine typically pulls the A/C system into vacuum during the last stage of recovery in order to extract as much refrigerant as possible from the system.

   Answer B is incorrect. Very little oil is recovered from the A/C system during a recovery process.

   Answer C is incorrect. Only Technician A is correct.

   Answer D is incorrect. Technician A is correct.

9. An automatic temperature control (ATC) A/C system displays a diagnostic trouble code (DTC) fault for the temperature blend door actuator motor. Technician A says the first step in the repair procedure is to replace the temperature blend door actuator. Technician B says that a diagnostic flow chart usually needs to be followed in order to find the problem. Who is correct?

TASK C.3.8

   A. A only
   B. B only
   C. Both A and B
   D. Neither A nor B

   Answer A is incorrect. The circuit components and the actuator should be tested prior to replacing the component.

   Answer B is correct. Only Technician B is correct. Diagnostic flow charts are logical troubleshooting steps used to find the root cause of vehicle problems.

   Answer C is incorrect. Only Technician B is correct.

   Answer D is incorrect. Technician B is correct.

10. An A/C thermal limiter switch is being diagnosed. Technician A says a thermal switch can be connected in series with the compressor clutch. Technician B says a thermal switch is usually mounted on the compressor. Who is correct?

TASK B.1.2

   A. A only
   B. B only
   C. Both A and B
   D. Neither A nor B

   Answer A is incorrect. Technician B is also correct.

   Answer B is incorrect. Technician A is also correct.

   Answer C is correct. Both Technicians are correct. Systems that use a thermal switch will sometimes connect this switch in series with the compressor and also will mount the switch on the compressor.

   Answer D is incorrect. Both Technicians are correct.

TASK B.1.6

11. Technician A says a replacement compressor should have the same type of line connection points as the old compressor. Technician B says the mounting brackets and other fasteners on a replacement compressor should be identical to those on the old compressor. Who is correct?

A. A only
B. B only
C. Both A and B
D. Neither A nor B

Answer A is incorrect. Technician B is also correct.

Answer B is incorrect. Technician A is also correct.

Answer C is correct. Both Technicians are correct. The line connection points as well as the mounting positions and brackets on the replacement compressor should be identical to the old compressor.

Answer D is incorrect. Both Technicians are correct.

TASK A.1

12. Technician A says if an air conditioning system has too much refrigerant oil, the performance of the system will suffer. Technician B says poor system performance can be caused by an overcharge of refrigerant in the system. Who is correct?

A. A only
B. B only
C. Both A and B
D. Neither A nor B

Answer A is incorrect. Technician B is also correct.

Answer B is incorrect. Technician A is also correct.

Answer C is correct. Both Technicians are correct. An A/C system will have slightly reduced performance if it has excessive refrigerant oil installed. An A/C system will have very poor performance if it is overcharged with refrigerant.

Answer D is incorrect. Both Technicians are correct.

TASK B.2.8

13. Technician A says that restricted refrigerant passages in the evaporator may cause frosting of the evaporator outlet pipe. Technician B says restricted refrigerant passages in the evaporator may cause much higher-than-specified low-side pressures. Who is correct?

A. A only
B. B only
C. Both A and B
D. Neither A nor B

Answer A is correct. Only Technician A is correct. A restricted evaporator core could cause frosting of the outlet pipe due to the refrigerant still taking on heat.

Answer B is incorrect. A restricted evaporator core would cause lower pressure in the low side of the A/C system.

Answer C is incorrect. Only Technician A is correct.

Answer D is incorrect. Technician A is correct.

14. A performance test reveals the pressure on the low side of the refrigerant system is higher than specification and the pressure on the high side of the refrigerant system is lower than specification. Technician A says that the A/C condenser could be restricted internally. Technician B says that the intake reed valve in the compressor could be broken. Who is correct?

TASK A.2

A. A only
B. B only
C. Both A and B
D. Neither A nor B

Answer A is incorrect. A restricted condenser would likely cause the high-side pressure to be elevated.

Answer B is correct. Only Technician B is correct. A broken compressor reed valve could cause the pressure to be unusual. The low-side pressure would be elevated and the high-side pressure would be reduced.

Answer C is incorrect. Only Technician B is correct.

Answer D is incorrect. Technician B is correct.

15. An A/C compressor is being diagnosed for a slipping front clutch drive plate. A voltage test at the compressor clutch coil shows 8.5 volts when the compressor is engaged. Technician A says that the A/C clutch relay could have high resistance on the load-side contacts. Technician B says that the A/C coil ground could be loose. Who is correct?

TASK C.1.1

A. A only
B. B only
C. Both A and B
D. Neither A nor B

Answer A is incorrect. Technician B is also correct.

Answer B is incorrect. Technician A is also correct.

Answer C is correct. Both Technicians are correct. High resistance on relay contacts could cause the compressor clutch coil to not receive the full system voltage. A loose A/C coil ground could cause the coil to not receive full system voltage. Voltage drop tests could be performed with the system energized to further pinpoint the problem.

Answer D is incorrect. Both Technicians are correct.

16. Any of these conditions should cause the electric cooling fan relay to energize the fan motor EXCEPT:

TASK C.1.6

A. Coolant temperature above 230°F
B. Check engine light blinking with engine running
C. An output test using a scan tool
D. A/C refrigerant pressure above 400 psi

Answer A is incorrect. Coolant temperature above 230°F would likely cause the cooling fan relay to energize the fan motor.

Answer B is correct. A blinking check engine light would not typically cause the cooling fan to come on.

Answer C is incorrect. Some output tests with a scan tool could cause the fan relay to energize the fan motor.

Answer D is incorrect. Refrigerant pressure of 400 psi would likely cause the cooling fan relay to energize the fan motor.

TASK C.2.1

17. Technician A says that some vacuum leaks in the HVAC system can be diagnosed with an ultrasonic leak detector. Technician B says that some mechanical HVAC switching systems can be diagnosed by listening to the doors hit the limits while moving the control head back and forth. Who is correct?

A. A only
B. B only
C. Both A and B
D. Neither A nor B

Answer A is incorrect. Technician B is also correct.

Answer B is incorrect. Technician A is also correct.

Answer C is correct. Both Technicians are correct. An ultrasonic leak detector can sometimes be used to listen for low-level noises created by vacuum leaks. Mechanical HVAC controls will usually create an audible sound when moving the control head back and forth.

Answer D is incorrect. Both Technicians are correct.

TASK B.2.2

18. The airflow at the condenser is severely restricted with dirt and mud. Technician A says that the duct temperature will be higher than normal on a hot day. Technician B says that system pressures will be elevated. Who is correct?

A. A only
B. B only
C. Both A and B
D. Neither A nor B

Answer A is incorrect. Technician B is also correct.

Answer B is incorrect. Technician A is also correct.

Answer C is correct. Both Technicians are correct. Duct temperature will be increased and the system pressures will be elevated if the compressor gets covered with dirt and mud.

Answer D is incorrect. Both Technicians are correct.

TASK C.1.4

19. Which device is used as the input to cause the A/C compressor clutch to disengage during a parallel parking maneuver?

A. Coolant temperature sensor
B. Throttle position sensor
C. Oxygen sensor
D. Power steering switch

Answer A is incorrect. The coolant sensor is used to sense engine temperature. Some engine control modules will deactivate the A/C compressor clutch if engine temperature rises to near overheating range.

Answer B is incorrect. The throttle position sensor is used to sense throttle angle. Some engine control modules will deactivate the A/C compressor clutch if the throttle angle reaches very high levels.

Answer C is incorrect. The oxygen sensor is used to provide feedback of oxygen content in the exhaust.

Answer D is correct. The power steering switch is used to provide power steering information to the engine control modules. Some engine control modules deactivate the A/C compressor clutch when this switch senses high pressure, such as when parallel parking.

20. While testing the A/C system, the low-side pressure is found to be higher than normal. Cooling the expansion valve remote bulb lowers the low-side pressure. Technician A says the expansion valve could be defective. Technician B says the remote bulb may be improperly secured to the evaporator outlet. Who is correct?

TASK B.2.5

A. A only
B. B only
C. Both A and B
D. Neither A nor B

Answer A is incorrect. Technician B is also correct.

Answer B is incorrect. Technician A is also correct.

Answer C is correct. Both Technicians are correct. Cooling the expansion valve is a diagnostic test to find out if the valve changes its operation when the temperature changes. The expansion valve could be defective or the problem could be the remote sensing bulb is not secured properly.

Answer D is incorrect. Both Technicians are correct.

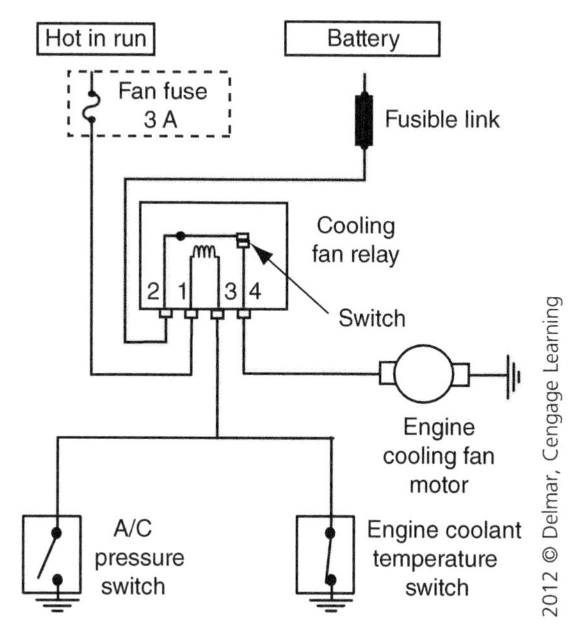

21. Referring to the figure above, the engine cooling fan does not operate when sitting in traffic for long periods of time but it will come on when the A/C is turned on. Technician A says that the fusible link could be open. Technician B says that the low-speed fan relay coil could be shorted. Who is correct?

TASK A.27

A. A only
B. B only
C. Both A and B
D. Neither A nor B

Answer A is incorrect. If the fusible link was open, the fan would never run.

Answer B is incorrect. The low-speed relay would have to be functional because the fan works when the A/C is turned on.

Answer C is incorrect. Neither Technician is correct.

Answer D is correct. Neither Technician is correct. The likely problem would be in the engine coolant temperature switch.

TASK A.23, A.24

22. The vacuum valve in the radiator cap is stuck closed. The result of this problem could be:

    A. Collapsed upper radiator hose after the engine was shut off
    B. Excessive cooling system pressure at normal engine temperature
    C. Engine overheating when operating under a heavy load
    D. Engine overheating during extended idle periods

Answer A is correct. The upper radiator hose would collapse due to vacuum being present in the radiator. The vacuum valve operates by allowing coolant to return to the radiator after the engine has cooled down.

Answer B is incorrect. A problem with the pressure release valve would not cause the cooling system to develop excessive pressure.

Answer C is incorrect. There are several possible causes for engine overheating but a vacuum valve on the radiator cap would not cause overheating.

Answer D is incorrect. An inoperative cooling fan is the likely cause of overheating during long idle times.

TASK C.3.9

23. An automatic temperature control (ATC) processor needs to be replaced. Technician A says that the new processor may need to be reprogrammed with a scan tool. Technician B says that the HVAC control head will have to be replaced at the same time. Who is correct?

    A. A only
    B. B only
    C. Both A and B
    D. Neither A nor B

Answer A is correct. Only Technician A is correct. Some ATC processors require the technician to reprogram the new component during a replacement procedure.

Answer B is incorrect. It is not common to have to replace the control head at the same time as the processor.

Answer C is incorrect. Only Technician A is correct.

Answer D is incorrect. Technician A is correct.

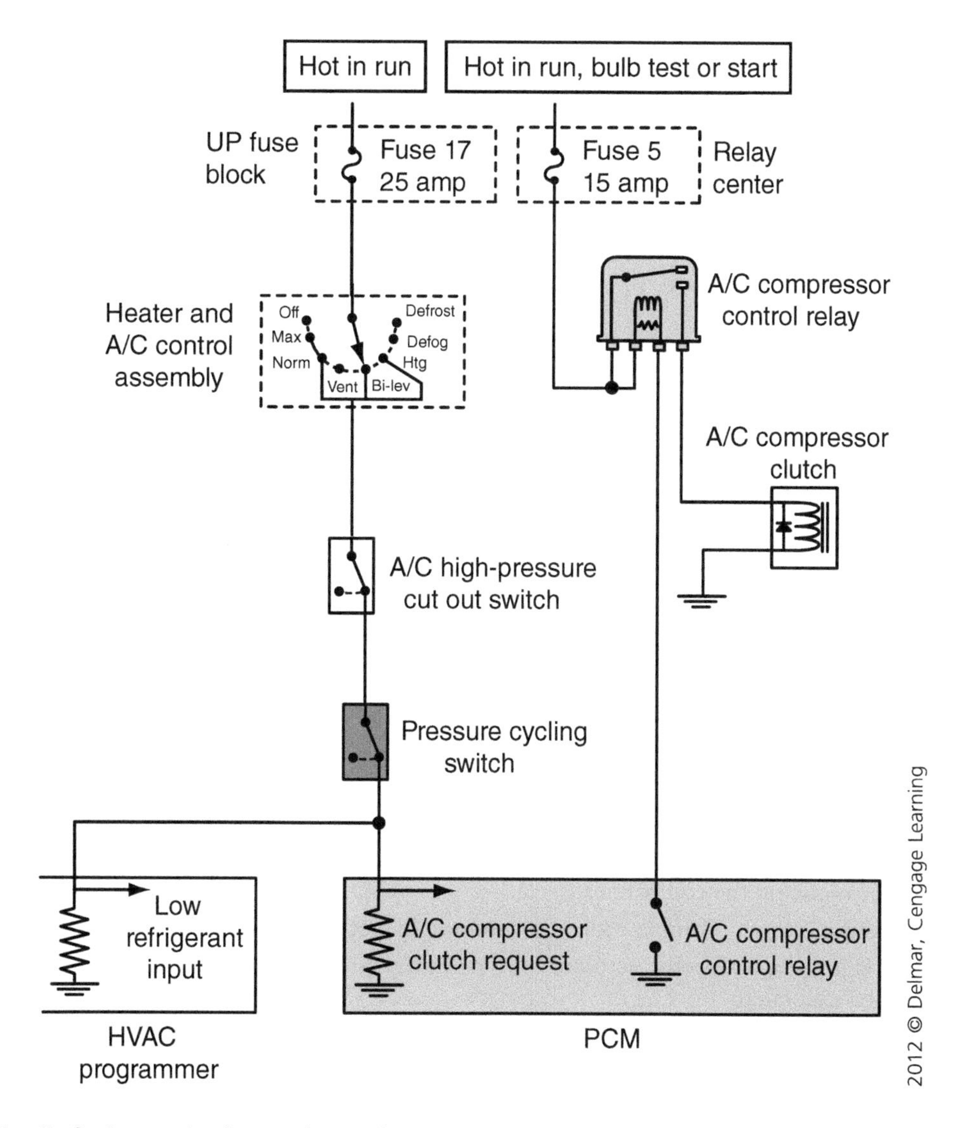

24. Referring to the figure above, the A/C compressor clutch will not engage when the A/C is selected at the HVAC control, but the compressor clutch will engage when commanding it on with a scan tool function output test. Technician A says that the HVAC control head could have an internal fault. Technician B says that the 15 amp fuse in the relay center could be open. Who is correct?

A. A only
B. B only
C. Both A and B
D. Neither A nor B

Answer A is correct. Only Technician A is correct. It is possible that the HVAC control head could cause the A/C not to engage.

Answer B is incorrect. Since the A/C compressor will engage using the function output test on the scan tool, the PCM, relay, 15 amp fuse, and clutch coil would not be the problem.

Answer C is incorrect. Only Technician A is correct.

Answer D is incorrect. Technician A is correct.

TASK A.22, A.28

25. The heater water valve is being checked. The hose between the control valve and heater core is hot, and the outlet hose from the heater core is much cooler. Technician A says the coolant control valve may be defective and should be cleaned or replaced. Technician B says the heater core may be clogged and should be cleaned or replaced. Who is correct?

    A. A only
    B. B only
    C. Both A and B
    D. Neither A nor B

Answer A is incorrect. The heater control valve is functioning since the heater hose entering the heater core is hot.

Answer B is correct. Only Technician B is correct. A clogged or restricted heater core could cause the heater outlet hose to be much cooler than the inlet.

Answer C is incorrect. Only Technician B is correct.

Answer D is incorrect. Technician B is correct.

TASK C.1.7, C.3.9

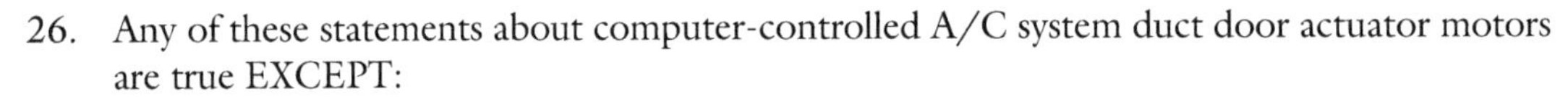

26. Any of these statements about computer-controlled A/C system duct door actuator motors are true EXCEPT:

    A. Some actuator motors are calibrated automatically using the self-diagnostic mode of the control head.
    B. AC system and component problems sometimes produce diagnostic trouble codes.
    C. The control rods must be calibrated manually on some systems.
    D. The actuator motor may require calibration after motor replacement.

Answer A is incorrect. Some electric actuators can be calibrated automatically in self-diagnostic mode on the control head.

Answer B is incorrect. Trouble codes are common in electronic air conditioning systems.

Answer C is correct. Electric actuator motors for the duct doors do not have control rods that can be calibrated.

Answer D is incorrect. Calibration of the electric actuators is sometimes necessary after they are replaced.

TASK A.7

27. Technician A says if the refrigerant is contaminated with moisture, chemical changes in the refrigerant can cause corrosion. Technician B says if refrigerant oil is allowed to mix with the refrigerant the system will not build up the necessary pressures to work efficiently. Who is correct?

    A. A only
    B. B only
    C. Both A and B
    D. Neither A nor B

Answer A is correct. Only Technician A is correct. If moisture enters the A/C system, it could create an acidic substance that would cause corrosion.

Answer B is incorrect. It is normal for refrigerant oil to mix with the refrigerant.

Answer C is incorrect. Only Technician A is correct.

Answer D is incorrect. Technician A is correct.

28. Technician A says that a blower resistor is used to lower voltage to the blower motor to create different speeds. Technician B says that the resistance provided by the blower resistor is highest at the lowest blower speed. Who is correct?

TASK C.1.2

A. A only
B. B only
C. Both A and B
D. Neither A nor B

Answer A is incorrect. Technician B is also correct.

Answer B is incorrect. Technician A is also correct.

Answer C is correct. Both Technicians are correct. Blower resistors are used to change the blower speeds by varying the different amount of circuit resistance depending on the switch setting. As more resistors are included in the blower circuit, the available voltage for the blower decreases. All of the blower resistors are in series with the blower motor when the switch is on low.

Answer D is incorrect. Both Technicians are correct.

29. The compressor clutch diode is being diagnosed. Technician A says a purpose of the diode is to control current flow through the A/C clutch coil when the coil is energized. Technician B says a purpose of the diode is to prevent spikes of high voltage created by the clutch coil operation from damaging delicate electronic components. Who is correct?

TASK C.1.3

A. A only
B. B only
C. Both A and B
D. Neither A nor B

Answer A is incorrect. The A/C clutch diode does not control current flow in the coil when it is energized. The diode does not conduct due to the reverse polarity when power is supplied to the coil.

Answer B is correct. Only Technician B is correct. The A/C clutch diode suppresses the voltage spike created when the coil is de-energized.

Answer C is incorrect. Only Technician B is correct.

Answer D is incorrect. Technician B is correct.

30. A car is equipped with an A/C compressor driven by a serpentine belt. As the compressor cycles, the belt squeals briefly. Technician A says the belt tensioner may be weak. Technician B says that the system may be undercharged with refrigerant. Who is correct?

TASK B.1.3

A. A only
B. B only
C. Both A and B
D. Neither A nor B

Answer A is correct. Only Technician A is correct. A weak belt tensioner can cause the belt to squeal when the A/C compressor engages.

Answer B is incorrect. An undercharged A/C system will not cause belt squeal. This condition can cause the system to cycle on and off more frequently.

Answer C is incorrect. Only Technician A is correct.

Answer D is incorrect. Technician A is correct.

TASK A.1

31. Any of these statements concerning the refrigeration system are true EXCEPT:
    A. Refrigerant leaves the compressor as a high-pressure, high-temperature vapor.
    B. Impurities are removed by the receiver/drier.
    C. The expansion valve controls the flow of refrigerant into the evaporator core.
    D. Warm air passing through the evaporator causes heat to be absorbed into the refrigerant.

Answer A is incorrect. Refrigerant enters the compressor as a low-pressure, low-temperature vapor and it leaves as a high-pressure, high-temperature vapor.

Answer B is correct. The receiver/drier does not remove the impurities from the refrigeration system.

Answer C is incorrect. The expansion valve acts as the metering device to control the flow of refrigerant into the evaporator core.

Answer D is incorrect. Warm air passes over the evaporator core and gives up its heat to the boiling refrigerant inside the evaporator core.

TASK A.14

32. A scan tool can be used for any of these actions EXCEPT:
    A. Retrieving diagnostic trouble codes (DTCs)
    B. Repairing a poor terminal connection at a sensor
    C. Performing output tests on actuators that are controlled by a control module
    D. Viewing live data from sensors that are inputs to a control module

Answer A is incorrect. Scan tools can retrieve DTCs from electronically controlled systems.

Answer B is correct. A scan tool does not have the ability to make a wiring repair.

Answer C is incorrect. Scan tools can perform output tests on actuators that are controlled by a control module.

Answer D is incorrect. Scan tools can display live data from input sensors to control modules.

TASK A.2

33. The liquid line has frost on it right at the point that exits the receiver/drier. Technician A says that the receiver/drier could be restricted. Technician B says that the compressor could have faulty reed valves. Who is correct?
    A. A only
    B. B only
    C. Both A and B
    D. Neither A nor B

Answer A is correct. Only Technician A is correct. Frost on the receiver/drier exit line would indicate a restricted receiver/drier. There should not be a drop in temperature at the receiver/drier.

Answer B is incorrect. Faulty compressor reed valves would cause the system pressures to be unusual. The high-side pressure would be lower than normal and the low-side pressure would be higher than normal.

Answer C is incorrect. Only Technician A is correct.

Answer D is incorrect. Technician A is correct.

34. The blower motor does not operate on high speed. All other speeds are operational. Technician A says the blower ground wire may be loose or broken. Technician B says the high-speed relay may be defective. Who is correct?

TASK C.1.2

A. A only
B. B only
C. Both A and B
D. Neither A nor B

Answer A is incorrect. A blower motor ground wire problem would cause problems at all speeds.

Answer B is correct. Only Technician B is correct. A defective blower motor relay can cause the high speed to be inoperative while having normal operation of the lower speeds.

Answer C is incorrect. Only Technician B is correct.

Answer D is incorrect. Technician B is correct.

35. An ambient temperature sensor is being diagnosed. Technician A says this device senses outside air temperature by varying the resistance as the outside light level changes. Technician B says that this device is typically a variable negative temperature coefficient (NTC) thermistor. Who is correct?

TASK C.3.5

A. A only
B. B only
C. Both A and B
D. Neither A nor B

Answer A is incorrect. Thermistors vary resistance as the temperature changes.

Answer B is correct. Only Technician B is correct. Outside air temperature sensors are typically NTC thermistors. These resistors operate by changing the resistance inversely according to the temperature.

Answer C is incorrect. Only Technician B is correct.

Answer D is incorrect. Technician B is correct.

36. Which of these test tools would be the most likely option to use when testing the resistance of a cabin temperature sensor?

TASK A.14

A. Digital multi-meter
B. Scan tool
C. Digital storage oscilloscope (DSO)
D. Jumper wire

Answer A is correct. A digital multi-meter is a valuable tool for testing voltage, current, and resistance in electronic circuits. It is accurate and safe to use on electronic circuits because of its high impedance.

Answer B is incorrect. A scan tool is not normally used to test the resistance of a sensor. These tools are commonly used to view live data, retrieve trouble codes, and perform output tests on electronically controlled systems.

Answer C is incorrect. A DSO is not normally used to test the resistance of a sensor. These tools are commonly used to view the waveforms of voltage and current within a circuit.

Answer D is incorrect. A jumper wire would not be used to test the resistance of a sensor. Jumper wires are sometimes used to by-pass parts of the circuit during the troubleshooting procedure.

TASK A.2

37. The low-side pressure is lower than normal. Which of these is the LEAST LIKELY cause?

A. A faulty metering device
B. Poor airflow across the evaporator
C. A restriction in the low side of the system
D. System overcharged with refrigerant

Answer A is incorrect. A restricted metering device could cause the low-side pressure to be lower than normal.

Answer B is incorrect. Poor airflow across the evaporator core could cause the low-side pressure to be lower than normal.

Answer C is incorrect. A low-side restriction could cause the low-side pressure to be lower than normal.

Answer D is correct. An overcharged A/C system would cause the A/C system pressures to be elevated, especially on the high side of the system.

TASK B.2.1

38. Quick-connect refrigerant lines are being serviced. For which process are special tools required?

A. Disconnecting spring lock couplings
B. Connecting spring lock couplings
C. Installing the o-rings
D. Inspecting the couplings

Answer A is correct. Quick-connect couplings require the use of a special tool to disconnect the lines.

Answer B is incorrect. No tools are needed to connect these types of fittings.

Answer C is incorrect. Installing the o-rings does not require special tools.

Answer D is incorrect. The coupling can be inspected without special tools.

TASK C.3.11

39. An automatic temperature control (ATC) problem is being diagnosed. The scan tool gives a fault of "data bus open." Technician A says that the problem could be a data wire rubbing a sharp body component. Technician B says that the problem could be a broken data wire. Who is correct?

A. A only
B. B only
C. Both A and B
D. Neither A nor B

Answer A is incorrect. A wire rubbing a sharp body component would likely cause a different trouble code.

Answer B is correct. Only Technician B is correct. A broken data wire could cause the fault "data bus open" due to no current flowing through an open wire.

Answer C is incorrect. Only Technician B is correct.

Answer D is incorrect. Technician B is correct.

40. The blower motor fuse blows after the blower is operated on high speed for several minutes. The fuse is replaced and blows again after the same amount of time with the blower on high. Technician A says that the problem is likely a direct short to ground before the load in blower circuit. Technician B says that the blower motor could have a tight bearing. Who is correct?

TASK C.1.1

A. A only
B. B only
C. Both A and B
D. Neither A nor B

Answer A is incorrect. A direct short to ground before the load would immediately blow the blower fuse.

Answer B is correct. Only Technician B is correct. A tight blower motor bearing could cause the blower fuse to blow after being operated for a few minutes due to the expansion of the bearing caused by excessive heat.

Answer C is incorrect. Only Technician B is correct.

Answer D is incorrect. Technician B is correct.

41. A vehicle with automatic temperature control (ATC) is malfunctioning. A scan tool is connected and a code is retrieved from the system that indicates a fault in the sun load sensor. Technician A says that the sensor should be tested with a test light. Technician B says the sensor can be checked with a digital ohmmeter to see if it varies resistance when a flashlight is shined onto it. Who is correct?

TASK C.3.1, C.3.3

A. A only
B. B only
C. Both A and B
D. Neither A nor B

Answer A is incorrect. A test light should never be used when testing electronic components due to its low resistance.

Answer B is correct. Only Technician B is correct. A sun load sensor could be tested with a digital ohmmeter to determine if it will vary its resistance as the light level changes.

Answer C is incorrect. Only Technician B is correct.

Answer D is incorrect. Technician B is correct.

42. Technician A says that the computer senses the position of the mode door by monitoring the signal voltage of the position sensor. Technician B says that the computer provides the ground for the position sensor. Who is correct?

TASK C.1.7

A. A only
B. B only
C. Both A and B
D. Neither A nor B

Answer A is incorrect. Technician B is also correct.

Answer B is incorrect. Technician A is also correct.

Answer C is correct. Both Technicians are correct. The HVAC computer uses a potentiometer to sense the position of the mode door. These sensors are composed of three wires. One wire is a 5 volt supply, another wire is a ground, while the third wire is the sensing wire.

Answer D is incorrect. Both Technicians are correct.

TASK B.2.1

43. Technician A says that the o-rings should be reused on most A/C repairs. Technician B says that the accumulation of oily residue around an A/C line connection is an indication of a leak in the system. Who is correct?

A. A only
B. B only
C. Both A and B
D. Neither A nor B

Answer A is incorrect. New o-rings should always be installed if possible when servicing A/C systems.

Answer B is correct. Only Technician B is correct. Oily residue around A/C line connections is typically an indication of a refrigerant leak.

Answer C is incorrect. Only Technician B is correct.

Answer D is incorrect. Technician B is correct.

TASK C.3.4

44. While testing a compressor clutch, a fused jumper wire is used to by-pass the load side of the A/C clutch relay and the compressor clutch engages. The most likely problem in this system would be:

A. An open compressor clutch coil
B. An open pressure cycling switch
C. An open compressor clutch coil ground circuit
D. An open relay power feed circuit

Answer A is incorrect. The compressor clutch would not engage if the coil was open.

Answer B is correct. An open pressure cycling switch can cause the relay to not receive a signal from the power train control module (PCM) to turn on the compressor.

Answer C is incorrect. The compressor clutch would not engage if the ground circuit was open.

Answer D is incorrect. The compressor clutch would not engage by by-passing the relay if the relay power feed was open.

TASK A.5, A.6, A.8, A.15

45. Technician A says it is common to mix R-12 and R-134a when servicing late-model A/C systems. Technician B says that some A/C service machines can recover, evacuate, and recharge A/C systems. Who is correct?

A. A only
B. B only
C. Both A and B
D. Neither A nor B

Answer A is incorrect. R-12 and R-134a should never be mixed.

Answer B is correct. Only Technician B is correct. It is common for A/C service machines to recover, evacuate, and recharge the A/C system.

Answer C is incorrect. Only Technician B is correct.

Answer D is incorrect. Technician B is correct.

46. Technician A says that the refrigerant should be recovered before connecting a refrigerant identifier. Technician B says that a refrigerant identifier can test for flammable refrigerant substances. Who is correct?

TASK A.5

A. A only
B. B only
C. Both A and B
D. Neither A nor B

Answer A is incorrect. The identifier should be connected prior to connecting the recovery equipment. Some recovery/recharging machines have a built-in identifier.

Answer B is correct. Only Technician B is correct. Refrigerant identifiers can test for flammable substances in the A/C system.

Answer C is incorrect. Only Technician B is correct.

Answer D is incorrect. Technician B is correct.

47. Technician A says that some data communication network wires can be repaired by carefully soldering the wires. Technician B says that some data network wiring failures require the use of a digital multi-meter to diagnose. Who is correct?

TASK C.3.11

A. A only
B. B only
C. Both A and B
D. Neither A nor B

Answer A is incorrect. Technician B is also correct.

Answer B is incorrect. Technician A is also correct.

Answer C is correct. Both Technicians are correct. Some data communication wires can be repaired using the solder method. In addition, the diagnosis of a data communication network requires the use of a digital multi-meter.

Answer D is incorrect. Both Technicians are correct.

48. A stuck open engine cooling system thermostat could cause any of these symptoms EXCEPT:

TASK A.25

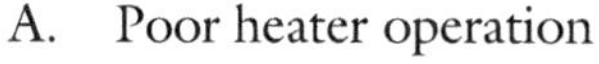

A. Poor heater operation
B. Below-normal engine operating temperature
C. Increased fuel consumption
D. Coolant loss

Answer A is incorrect. Poor heater operation could be caused by a stuck open thermostat.

Answer B is incorrect. A stuck open thermostat would cause the engine to run cooler thannormal.

Answer C is incorrect. The engine would use more fuel because of not being able to reach and maintain normal operating temperature.

Answer D is correct. A stuck open thermostat would not typically cause coolant loss.

TASK C.2.5

49. A vehicle with dual-zone climate control is being diagnosed. Technician A says that these systems use two separate duct boxes to deliver the varied airflow to the driver and passenger. Technician B says that these systems use two separate cabin air temperature sensors to provide feedback from each side of the car. Who is correct?

    A. A only
    B. B only
    C. Both A and B
    D. Neither A nor B

    Answer A is incorrect. A dual-zone climate control system does not use two separate duct boxes. A complex duct box with two blend air doors is typically the way that dual-zone systems are designed.

    Answer B is correct. Only Technician B is correct. An electronic dual-zone HVAC system would have a cabin air temperature sensor for both sides of the cabin to sense driver- and passenger-side temperature.

    Answer C is incorrect. Only Technician B is correct.

    Answer D is incorrect. Technician B is correct.

TASK B.1.4

50. Technician A says that the compressor clutch coil can be tested with a digital ohmmeter. Technician B says that some compressor clutch coils have a diode wired in parallel with the coil. Who is correct?

    A. A only
    B. B only
    C. Both A and B
    D. Neither A nor B

    Answer A is incorrect. Technician B is also correct.

    Answer B is incorrect. Technician A is also correct.

    Answer C is correct. Both Technicians are correct. A digital ohmmeter is the recommended tool for testing an A/C clutch coil. Some compressors have a diode wired in parallel with the coil to handle the voltage spike when the coil is de-energized.

    Answer D is incorrect. Both Technicians are correct.

PREPARATION EXAM 6—ANSWER KEY

1. A
2. A
3. D
4. C
5. C
6. C
7. C
8. C
9. C
10. D
11. C
12. A
13. C
14. B
15. A
16. C
17. B
18. D
19. B
20. B
21. D
22. B
23. C
24. C
25. B
26. C
27. B
28. B
29. A
30. D
31. C
32. C
33. B
34. A
35. D
36. A
37. C
38. D
39. A
40. B
41. C
42. C
43. B
44. B
45. C
46. A
47. B
48. C
49. C
50. C

PREPARATION EXAM 6—EXPLANATIONS

1. Technician A says that the refrigerant enters the compressor as a low-pressure gas. Technician B says that the refrigerant leaves the compressor as a high-pressure liquid. Who is correct?

TASK A.1

A. A only
B. B only
C. Both A and B
D. Neither A nor B

Answer A is correct. Only Technician A is correct. The A/C compressor pulls in low-pressure gas when the system is operating.

Answer B is incorrect. The compressor discharges high-pressure gas when the system is operating normally.

Answer C is incorrect. Only Technician A is correct.

Answer D is incorrect. Technician A is correct.

TASK A.20, A.28

2. Which of these statements is LEAST LIKELY to cause heating system problems?

A. High concentration of antifreeze in the coolant
B. Sticking or disconnected blend door
C. Heater control valves stuck open
D. Air pockets in the heater core

Answer A is correct. Having a high concentration of antifreeze in the coolant would not cause a heating system problem.

Answer B is incorrect. A sticking or disconnected blend air door could cause the heating system to not function right.

Answer C is incorrect. A heater control valve stuck open would not cause a problem with the heater performance.

Answer D is incorrect. An air pocket in the heater core could cause the heater system to not perform well.

TASK A.3

3. All of following characteristics of a normal operating A/C system are correct EXCEPT:

A. The discharge line should be warm or hot when the A/C system is engaged.
B. The liquid line should be warm or hot when the A/C system is engaged.
C. The HVAC drain tube should produce water after several minutes of operation in warm conditions.
D. The components on the low side should be frosty when the A/C system is engaged.

Answer A is incorrect. The discharge line should be warm or hot when the A/C system is engaged.

Answer B is incorrect. The liquid line should be warm or hot when the A/C system is engaged.

Answer C is incorrect. The HVAC drain tube should normally produce water after several minutes of operation in warm conditions. This water originates on the fins of the evaporator after it is pulled from the air.

Answer D is correct. The A/C system should never produce frost when the system is engaged.

TASK C.1.3

4. The A/C compressor clutch relay is being diagnosed. Technician A says that the relay coil can be tested for an open with a digital ohmmeter. Technician B says that some A/C compressor clutch relays can be activated with an output test using a scan tool. Who is correct?

A. A only
B. B only
C. Both A and B
D. Neither A nor B

Answer A is incorrect. Technician B is also correct.

Answer B is incorrect. Technician A is also correct.

Answer C is correct. Both Technicians are correct. Relays can be bench tested with a digital ohmmeter. The coil can be tested for the correct resistance value as well as the normally open and normally closed contacts. Some A/C compressor clutch relays can be activated with the scan tool output test.

Answer D is incorrect. Both Technicians are correct.

5. Any of these statements about the tools needed to service the refrigerant system are true EXCEPT:

TASK A.5

   A. Separate gauges and other refrigerant-handling equipment should be used for different types of refrigerant.
   B. Manifold gauge sets for R-134a can be identified by unique service fitting connections.
   C. R-134a pressure gauges are much stronger than R-12 gauges.
   D. Proper identification of service equipment and hoses is important to prevent cross-contamination of refrigerant.

   Answer A is incorrect. A repair shop should have separate gauges and handling equipment for each type of refrigerant serviced.

   Answer B is incorrect. R-134a manifold service fittings are unlike R-12 service fittings. R-134a service fitting are the quick-disconnect style.

   Answer C is correct. R-134a pressure gauges are not any stronger than R-12 gauges.

   Answer D is incorrect. Each type of refrigerant service equipment should be easily identifiable in order to prevent cross-contamination of the refrigerant.

6. Technician A says that the dual pressure switch prevents the compressor from operating if the system has lost all of the refrigerant. Technician B says that the dual pressure switch can cause the compressor to turn off if the high-side pressure exceeds specifications. Who is correct?

TASK B.1.2, C.1.5

   A. A only
   B. B only
   C. Both A and B
   D. Neither A nor B

   Answer A is incorrect. Technician B is also correct.

   Answer B is incorrect. Technician A is also correct.

   Answer C is correct. Both Technicians are correct. The dual pressure switch opens if pressure drops very low, and it also opens if pressure rises above specified ranges.

   Answer D is incorrect. Both Technicians are correct.

7. An A/C system performance test is performed on a late-model car. Technician A inserts a thermometer into the air duct at the center of the dash and monitors discharge temperature. Technician B connects a manifold pressure gauge set to the service fittings to monitor pressure. Who is correct?

TASK A.1

   A. A only
   B. B only
   C. Both A and B
   D. Neither A nor B

   Answer A is incorrect. Technician B is also correct.

   Answer B is incorrect. Technician A is also correct.

   Answer C is correct. Both Technicians are correct. The duct temperature and the system pressures should be checked during an A/C system performance test. The duct temperature should be between 40°F to 50°F (4.4°C–10°C), depending on the ambient temperature and humidity levels. The system pressures will also vary depending on the ambient temperature and humidity levels.

   Answer D is incorrect. Both Technicians are correct.

TASK A.2

8. The suction line is covered with a thick frost. Technician A says that this might indicate that the expansion valve is flooding the evaporator. Technician B says that the evaporator core may have a blockage. Who is correct?

A. A only
B. B only
C. Both A and B
D. Neither A nor B

Answer A is incorrect. Technician B is also correct.

Answer B is incorrect. Technician A is also correct.

Answer C is correct. Both Technicians are correct. Frost on the suction line could be caused by an expansion valve stuck open and flooding the evaporator. In addition, frost could be caused by a blocked evaporator core.

Answer D is incorrect. Both Technicians are correct.

TASK A.7

9. A refrigerant identifier will detect any of these contaminants in an A/C system EXCEPT:

A. Air
B. Mixed refrigerants
C. Sealer additive
D. Flammable substance

Answer A is incorrect. Refrigerant identifiers will detect the percentage of air in an A/C system. Refrigerant recovery/recharging machines will typically remove small quantities of air during the recovery and recharge process.

Answer B is incorrect. Refrigerant identifiers will detect mixed refrigerants in an A/C system. If a system has a mixed refrigerant, the normal shop equipment should not be connected to the system.

Answer C is correct. Refrigerant identifiers will not typically detect refrigerant sealer. A separate flow rate test is required to detect if sealer has been added to an A/C system.

Answer D is incorrect. Refrigerant identifiers will detect flammable substances. If a system tests positive for flammable refrigerant, the vehicle should be handled with care due to the possibility of fire.

TASK C.1.8, C.3.9

10. Technician A says that individual parts of an electronic HVAC control panel can be tested and replaced. Technician B says that electronic HVAC control panels should be tested with a test light for correct operation. Who is correct?

A. A only
B. B only
C. Both A and B
D. Neither A nor B

Answer A is incorrect. Electronic HVAC control panels do not typically have replaceable parts.

Answer B is incorrect. Test lights should not be used to test electronic HVAC control panels. Digital multi-meters should be used when testing any electronic components and circuitry.

Answer C is incorrect. Neither Technician is correct.

Answer D is correct. Neither Technician is correct. Electronic HVAC control panels do not have replaceable parts and should not be tested using a test light.

11. Technician A says a container of PAG refrigerant oil must be kept closed when not in use to prevent the oil from absorbing moisture. Technician B says that any time an A/C system has a component replaced, oil must be added. Who is correct?

TASK A.9

A. A only
B. B only
C. Both A and B
D. Neither A nor B

Answer A is incorrect. Technician B is also correct.

Answer B is incorrect. Technician A is also correct.

Answer C is correct. Both Technicians are correct. PAG oil is very hydroscopic, so it should be kept closed tightly when not in use. It is a good practice to purchase refrigerant oil in small containers in order to not waste much of the oil by letting it absorb moisture. Refrigerant oil should be added to each component as it is replaced in order to spread out the oil to various parts of the system.

Answer D is incorrect. Both Technicians are correct.

12. Any of these statements are correct about cabin air filter replacement EXCEPT:

TASK B.2.1

A. A clogged filter will not produce any noticeable problem in the HVAC system.
B. The filter reduces the amount of dust and pollen from the passenger compartment.
C. Driving conditions and terrain will alter the service intervals for the filter.
D. A visual inspection is required to evaluate the need for replacement of the filter.

Answer A is correct. A clogged cabin air filter would cause reduced airflow to all of the ducts of the HVAC system.

Answer B is incorrect. The cabin air filter reduces the amount of dust and pollen from the passenger compartment by trapping these substances inside the layers of the filter.

Answer C is incorrect. Driving conditions and terrain will alter the service intervals for the cabin air filter.

Answer D is incorrect. The cabin air filter should be inspected to evaluate the condition of the filter to determine if it needs to be replaced.

13. An evaporator core replacement is being performed on a late-model vehicle. Technician A disconnects the negative battery cable before working around the airbag components. Technician B stores the airbag components "face up" in a safe area while they are removed from the vehicle. Who is correct?

TASK A.11

A. A only
B. B only
C. Both A and B
D. Neither A nor B

Answer A is incorrect. Technician B is also correct.

Answer B is incorrect. Technician A is also correct.

Answer C is correct. Both Technicians are correct. The negative battery cable should be disconnected prior to working around airbag components. It is also advisable to wait about 10 minutes before beginning any work in order to let the capacitors in the airbag module discharge. It is also wise to store airbag components "face up" in a safe area while they are removed from the vehicle.

Answer D is incorrect. Both Technicians are correct.

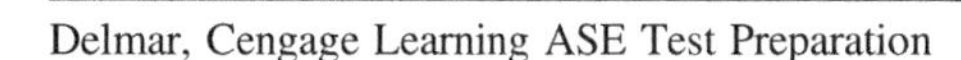

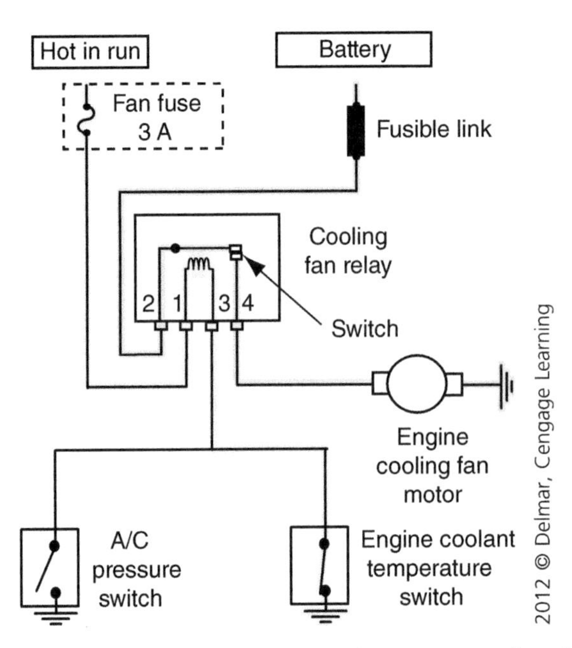

TASK A.27

14. Referring to the figure above, the engine cooling fan runs constantly with the key in the run position. Technician A says that the A/C pressure switch could stuck be open. Technician B says that the low-speed fan relay load-side contacts could be shorted. Who is correct?

A. A only
B. B only
C. Both A and B
D. Neither A nor B

Answer A is incorrect. An open A/C pressure switch would not cause the engine cooling fan to run constantly. A "stuck closed" A/C pressure switch could cause the fan to run constantly with the key in the run position.

Answer B is correct. Only Technician B is correct. Shorted fan relay contacts could cause the fan to run constantly. Since the load side of the relay receives power directly from the battery, the fan would likely cause the battery to run down as the fan would run constantly.

Answer C is incorrect. Only Technician B is correct.

Answer D is incorrect. Technician B is correct.

TASK B.1.3

15. The serpentine drive belt is being inspected during an A/C repair. The inside of the belt is cracked and has chunks missing every two inches. Technician A says that the belt should be replaced. Technician B says that the belt tensioner should be replaced every time the belt is replaced. Who is correct?

A. A only
B. B only
C. Both A and B
D. Neither A nor B

Answer A is correct. Only Technician A is correct. A belt in the described conditions should be replaced. The belt tensioner should be closely inspected during the belt replacement.

Answer B is incorrect. It is not necessary to replace the belt tensioner every time the belt is replaced. The tensioner should be closely inspected when the belt is replaced or inspected.

Answer C is incorrect. Only Technician A is correct.

Answer D is incorrect. Technician A is correct.

16. Technician A says that the ATC control head can be used on some vehicles to retrieve diagnostic trouble codes from the HVAC computer. Technician B says that an electronic scan tool can be used to retrieve ATC diagnostic trouble codes from a vehicle. Who is correct?

TASK C.3.9

A. A only
B. B only
C. Both A and B
D. Neither A nor B

Answer A is correct. An electronic scan tool can be used to retrieve diagnostic trouble codes from the ATC system.

Answer B is incorrect. Some ATC systems also allow diagnostic trouble codes to be displayed on the display of the control assembly. Codes from other systems like the engine and transmission are not normally available at the ATC control head.

Answer C is incorrect. Only Technician A is correct.

Answer D is incorrect. Technician A is correct.

17. Any of these statements about refrigerant oil are correct EXCEPT:

TASK B.1.5

A. PAG oil is the recommended lubricant for R-134a A/C systems.
B. The system oil level can be checked with a dipstick on late-model A/C systems.
C. Mineral oil is the recommended lubricant for R-12 A/C systems.
D. PAG oil is a synthetic substance.

Answer A is incorrect. PAG oil is the recommended oil for R-134a A/C systems. This oil comes in several viscosities and care should be taken to use the correct one.

Answer B is correct. Late-model compressors do not have a dipstick available for checking the system oil level.

Answer C is incorrect. R-12 A/C systems use mineral-based refrigerant oil.

Answer D is incorrect. PAG oil is a synthetic lubricant. This oil comes in several viscosities and care should be taken to use the correct one.

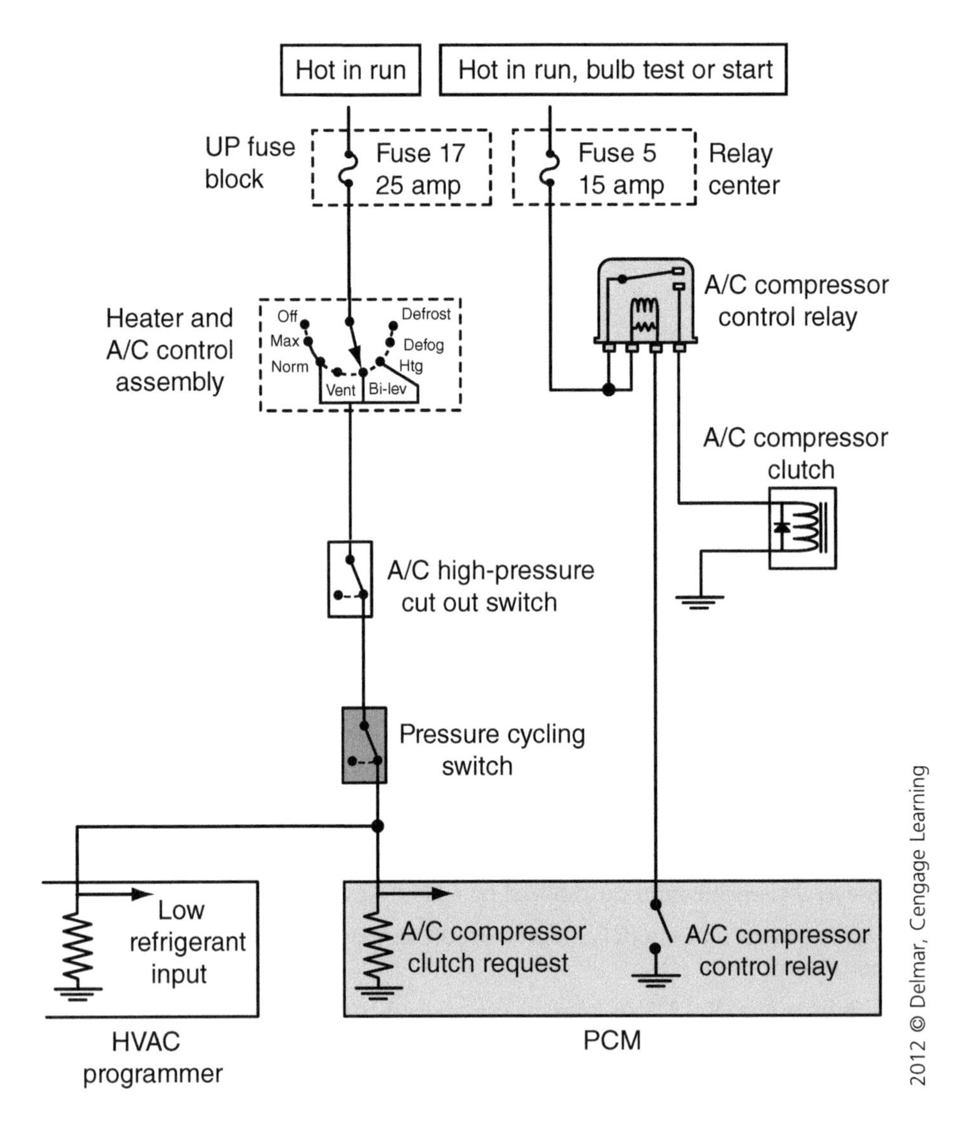

TASK B.1.1, B.1.2, B.1.4

18. Referring to the figure above, the A/C compressor clutch will not engage when the A/C is selected on the HVAC control head. The technician attempts to engage the compressor clutch using a scan tool output test but it still does not work. Technician A says that the high-pressure cutout switch could be faulty. Technician B says that the A/C clutch diode could be open. Who is correct?

A. A only
B. B only
C. Both A and B
D. Neither A nor B

Answer A is incorrect. The compressor would still have been energized with the scan tool even if the pressure cutout switch was open.

Answer B is incorrect. An open A/C clutch diode would not cause the problem described. The diode is just used to handle the voltage spike that occurs when the compressor clutch coil is de-energized.

Answer C is incorrect. Neither Technician is correct.

Answer D is correct. Neither Technician is correct. The problem in this circuit could be the fuse 5, the A/C compressor control relay, the compressor clutch, or possibly the PCM.

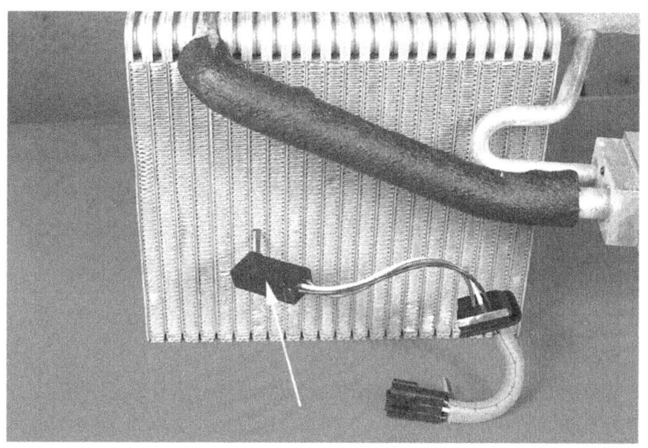

19. The component with the wire in the figure above could be any of these components EXCEPT:

TASK B.1.2, B.2.8

A. Evaporator temperature sensor
B. Thermal expansion valve
C. Thermostatic switch
D. Fin temperature probe

Answer A is incorrect. The item in the figure could be an evaporator temperature sensor since it is installed into the evaporator and it has three wires attached to it.

Answer B is correct. Thermal expansion valves do not have wires. Thermal expansion valves also have refrigerant lines connected to them.

Answer C is incorrect. The item in the figure could be a thermostatic switch since it is installed into the evaporator core and it has three wires connected to it.

Answer D is incorrect. The item in the figure could be a fin temperature probe since it is installed into the fins of the evaporator and it has three wires connected to it.

20. The A/C suction/discharge hose is being replaced on a late-model vehicle. Technician A says that o-rings used on the connections should be installed dry and clean. Technician B says that the metal sealing washers should be installed dry and clean. Who is correct?

TASK B.2.1

A. A only
B. B only
C. Both A and B
D. Neither A nor B

Answer A is incorrect. The o-rings should be installed with a small amount of mineral-type refrigerant oil applied to them when servicing the suction/discharge hose. The oil assists the o-rings in sealing well.

Answer B is correct. Only Technician B is correct. Metal sealing washers should be installed dry and clean when servicing the suction/discharge hose.

Answer C is incorrect. Only Technician B is correct.

Answer D is incorrect. Technician B is correct.

TASK B.2.2

21. Which of these problems would be the most likely result of a condenser that is covered with debris?

    A. The discharge line would be warm to the touch.
    B. The suction line would have condensation dripping off of it.
    C. The accumulator would have condensation dripping off of it.
    D. The high-side pressure would be elevated.

Answer A is incorrect. A very warm discharge line is a normal characteristic.

Answer B is incorrect. Condensation dripping off of the suction line is a normal characteristic.

Answer C is incorrect. Condensation dripping off of the accumulator is a normal characteristic.

Answer D is correct. The high-side pressure would be elevated if the condenser is covered with debris due to the reduced heat transfer capacity.

TASK B.2.6

22. An orifice tube is found to have heavy debris lodged on the screen. Technician A says that the screen can be cleaned and the orifice tube can be reused. Technician B says that the A/C system will need to be flushed and inspected to assure that all of the debris is removed from the system. Who is correct?

    A. A only
    B. B only
    C. Both A and B
    D. Neither A nor B

Answer A is incorrect. The orifice tube should be changed any time that it is exposed to inspect.

Answer B is correct. Only Technician B is correct. The system will need to be flushed when debris is spread throughout the system.

Answer C is incorrect. Only Technician B is correct.

Answer D is incorrect. Technician B is correct.

TASK B.2.4

23. A receiver/drier is being diagnosed. Technician A says the receiver/drier should be changed if the outlet is colder than the inlet. Technician B says the receiver/drier should be changed if the refrigerant system has been exposed to the atmosphere for a long period of time. Who is correct?

    A. A only
    B. B only
    C. Both A and B
    D. Neither A nor B

Answer A is incorrect. Technician B is also correct.

Answer B is incorrect. Technician A is also correct.

Answer C is correct. Both Technicians are correct. The receiver/drier should be changed if the outlet is cooler than the inlet. The receiver/drier should also be changed if the refrigerant system is exposed to the atmosphere for a long period of time due to the device getting saturated with humidity.

Answer D is incorrect. Both Technicians are correct.

24. The evaporator housing water drain is clogged. Technician A says that this may cause water to leak onto the front floorboard of the vehicle. Technician B says that this may cause water droplets to be present on the A/C duct vents when the system is operated. Who is correct?

TASK B.2.9

A. A only
B. B only
C. Both A and B
D. Neither A nor B

Answer A is incorrect. Technician B is also correct.

Answer B is incorrect. Technician A is also correct.

Answer C is correct. Both Technicians are correct. A clogged evaporator housing drain can cause water to leak onto the carpet in the front floorboard. This condition could also cause water droplet to be sent to the A/C duct vents as the system is operated.

Answer D is incorrect. Both Technicians are correct.

25. Technician A says that the coolant freeze protection can be checked with a voltmeter. Technician B says that coolant freeze protection can be checked with a refractometer. Who is correct?

TASK A.22

A. A only
B. B only
C. Both A and B
D. Neither A nor B

Answer A is incorrect. A voltmeter can be used to test the coolant for static and dynamic electrolysis. Static voltage can occur if the coolant mixture becomes too acidic. Dynamic voltage can occur if an electrical accessory has a poor ground.

Answer B is correct. Only Technician B is correct. Using a refractometer is the most accurate way to measure the freeze protection of the coolant.

Answer C is incorrect. Only Technician B is correct.

Answer D is incorrect. Technician B is correct.

26. The temperature is measured at the heater ducts with the HVAC control set at full heat after the vehicle has been run for 20 minutes. The duct temperature is found to be only 90°F. Technician A says that the engine thermostat should be checked for correct operation. Technician B says that the heater core may be restricted. Who is correct?

TASK A.25, A.29

A. A only
B. B only
C. Both A and B
D. Neither A nor B

Answer A is incorrect. Technician B is also correct.

Answer B is incorrect. Technician A is also correct.

Answer C is correct. Both Technicians are correct. A defective thermostat could cause the heater to not produce enough heat. A restricted heater core could cause the heater to not produce enough heat.

Answer D is incorrect. Both Technicians are correct.

TASK A.2

27. A vehicle has an A/C compressor that only runs for a few seconds at a time and then shuts off. Technician A says that the refrigerant may have too much refrigerant oil. Technician B says the system may be low on refrigerant. Who is correct?

A. A only

B. B only

C. Both A and B

D. Neither A nor B

Answer A is incorrect. Excessive refrigerant oil would not normally cause the A/C compressor to "short cycle."

Answer B is correct. Only Technician B is correct. A low refrigerant charge will often cause the A/C compressor to "short cycle."

Answer C is incorrect. Only Technician B is correct.

Answer D is incorrect. Technician B is correct.

TASK C.1.1

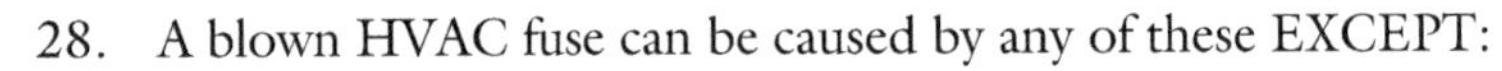

28. A blown HVAC fuse can be caused by any of these EXCEPT:

A. A short to ground in the temperature blend actuator circuit

B. A damaged cabin air temperature sensor

C. A damaged wiring harness

D. A short to ground in blower motor circuit

Answer A is incorrect. A short to ground in the temperature blend actuator circuit could cause a blown HVAC fuse by having excessive current.

Answer B is correct. A damaged cabin air temperature sensor would not cause the HVAC fuse to blow.

Answer C is incorrect. A damaged wiring harness could cause some type of electrical short circuit, which could cause the HVAC fuse to blow.

Answer D is incorrect. A short to ground in the blower circuit could cause the HVAC fuse to blow by increasing the current flow.

TASK A.4

29. A vehicle comes to a repair shop with an A/C complaint and the A/C system is found to be empty. Technician A says that nitrogen could be used to leak test the empty A/C system. Technician B says that an electronic leak detector could be used to leak test the empty A/C system. Who is correct?

A. A only

B. B only

C. Both A and B

D. Neither A nor B

Answer A is correct. Only Technician A is correct. Nitrogen is sometimes used to leak test the A/C system. The pressure can be monitored to sense leaks in the system or the system can be sprayed with a soapy solution to inspect for potential leaks.

Answer B is incorrect. An electronic leak detector will not work on an empty system.

Answer C is incorrect. Only Technician A is correct.

Answer D is incorrect. Technician A is correct.

30. An A/C system is being evacuated after a compressor replacement. Technician A says that this will help remove the foreign debris from the A/C system. Technician B says that the evacuation should be run for at least 10 minutes. Who is correct?

TASK A.6

A. A only
B. B only
C. Both A and B
D. Neither A nor B

Answer A is incorrect. Evacuation does not remove foreign debris from the A/C system.

Answer B is incorrect. The A/C system should be evacuated for at least 30 minutes if the system has been exposed to the atmosphere.

Answer C is incorrect. Neither Technician is correct.

Answer D is correct. Neither Technician is correct. Evacuation of the A/C system helps remove moisture from the A/C system and it should be performed for at least 30 minutes if the system has been exposed to the atmosphere.

31. The engine temperature sensor is being diagnosed. Technician A says that the A/C compressor may be deactivated if the engine temperature rises too high. Technician B says that the engine temperature sensor may cause the check engine light to illuminate if it develops a problem. Who is correct?

TASK C.1.4

A. A only
B. B only
C. Both A and B
D. Neither A nor B

Answer A is incorrect. Technician B is also correct.

Answer B is incorrect. Technician A is also correct.

Answer C is correct. Both Technicians are correct. Some vehicles will deactivate the A/C compressor if the engine temperature gets too high. This helps control engine temperature since the condenser would cool off if the compressor is turned off. In addition, engine temperature sensor faults could cause the check engine light to illuminate.

Answer D is incorrect. Both Technicians are correct.

32. A vehicle that uses electric HVAC duct actuators is being diagnosed for an HVAC air handling problem. The air from the ducts does not come out at the correct location. Technician A says that a digital voltmeter could be used to measure the position sensor voltage on the mode actuators. Technician B says that the mode doors may need to be calibrated with a scan tool. Who is correct?

TASK C.1.7, C.3.10

A. A only
B. B only
C. Both A and B
D. Neither A nor B

Answer A is incorrect. Technician B is also correct.

Answer B is incorrect. Technician A is also correct.

Answer C is correct. Both Technicians are correct. A digital voltmeter could be used to test the position sensor voltage on the mode actuators as well as to test the voltage to the motors. Electric actuators sometimes need to be calibrated with a scan tool.

Answer D is incorrect. Both Technicians are correct.

TASK C.1.6

33. What is the LEAST LIKELY cause of an inoperative condenser fan motor?

    A. A blown fan maxi-fuse
    B. A faulty engine thermostat
    C. A faulty engine control module
    D. A faulty A/C pressure sensor

Answer A is incorrect. A blown fan maxi-fuse could cause an inoperative condenser fan motor.

Answer B is correct. The engine thermostat would not cause an inoperative condenser fan motor.

Answer C is incorrect. A faulty engine control module could cause an inoperative condenser fan motor.

Answer D is incorrect. A faulty A/C pressure sensor could cause the condenser fan motor to not operate correctly.

TASK C.1.2

34. The blower motor operates slower than normal at all speed settings. A voltage test is performed at the blower connector with the switch in the high-speed switch position and 12.8 volts is measured. Technician A says that the cause could be a bad blower motor. Technician B says the problem could be a high resistance at the blower relay. Who is correct?

    A. A only
    B. B only
    C. Both A and B
    D. Neither A nor B

Answer A is correct. Only Technician A is correct. The voltage test shows that the blower motor is getting 12.8 volts. If the blower motor is operating slower than normal, then the cause could be a bad blower motor.

Answer B is incorrect. The blower motor would not be dropping 12.8 volts if the relay was faulty.

Answer C is incorrect. Only Technician A is correct.

Answer D is incorrect. Technician A is correct.

TASK A.8

35. An A/C system needs to be recharged after a condenser replacement. Technician A says that the A/C system can be charged accurately by watching the gauge pressures. Technician B says that the A/C system can be charged accurately by watching the duct temperatures. Who is correct?

    A. A only
    B. B only
    C. Both A and B
    D. Neither A nor B

Answer A is incorrect. An accurate charge is not possible by just watching the gauge pressures.

Answer B is incorrect. An accurate charge is not possible by just watching the duct temperatures.

Answer C is incorrect. Neither Technician is correct.

Answer D is correct. Neither Technician is correct. The exact weight should be charged into each A/C system in order to assure a quality repair. This can be accomplished by using an all-in-one machine or by using electronic scales during the charging process.

36. Technician A says that a dual-zone climate control system uses two blend doors. Technician B says that a dual-zone climate control system uses two fresh air actuators. Who is correct?

TASK C.1.7

A. A only
B. B only
C. Both A and B
D. Neither A nor B

Answer A is correct. Only Technician A is correct. Dual-zone climate control systems use a blend door for each side of the vehicle.

Answer B is incorrect. Dual-zone climate control systems do not typically use two fresh air actuators. These systems usually use two blend air doors.

Answer C is incorrect. Only Technician A is correct.

Answer D is incorrect. Technician A is correct.

37. A vacuum-controlled HVAC system is being diagnosed. The air only comes out of the defrost duct, but the temperature and blower speeds operate normally. Technician A says that a kinked vacuum hose may be the cause. Technician B says that a broken vacuum hose may be the cause. Who is correct?

TASK C.2.1, C.2.4

A. A only
B. B only
C. Both A and B
D. Neither A nor B

Answer A is incorrect. Technician B is also correct.

Answer B is incorrect. Technician A is also correct.

Answer C is correct. Both Technicians are correct. A kinked vacuum hose or broken vacuum hose could cause the HVAC system to be stuck in defrost mode.

Answer D is incorrect. Both Technicians are correct.

38. Any of these are common methods of moving the HVAC duct doors EXCEPT:

TASK C.2.1, C.2.3

A. Electric actuators
B. Vacuum actuators
C. Cables
D. Air pressure

Answer A is incorrect. Electric actuators are often used to move the HVAC duct doors.

Answer B is incorrect. Vacuum actuators are often used to move the HVAC duct doors.

Answer C is incorrect. Cables are sometimes used to move the HVAC duct doors.

Answer D is correct. Cars and light-duty trucks do not use air pressure to move the HVAC duct doors.

TASK C.2.5

39. An HVAC duct door is being replaced on a late-model vehicle. Technician A says that the negative battery cable should be disconnected prior to beginning this repair. Technician B says that this repair can typically be made without opening up the duct box. Who is correct?

A. A only
B. B only
C. Both A and B
D. Neither A nor B

Answer A is correct. Only Technician A is correct. The negative battery cable would need to be disconnected prior to beginning to disassemble the dash panel. Many types of vehicles would require the whole HVAC duct box to be removed to replace a duct door.

Answer B is incorrect. Replacing an HVAC duct door will typically require the whole HVAC duct box to be removed from the vehicle.

Answer C is incorrect. Only Technician A is correct.

Answer D is incorrect. Technician A is correct.

TASK C.3.3

40. A vehicle with electronic dual-zone climate control has a problem of the driver's side only blowing cool air no matter what the driver setting is adjusted to. Technician A says that the driver's mode door actuator could be defective. Technician B says that the driver's blend door actuator could be defective. Who is correct?

A. A only
B. B only
C. Both A and B
D. Neither A nor B

Answer A is incorrect. The mode actuator will only affect the location that the air is discharged. It will not cause the temperature of the air to be too cool.

Answer B is correct. Only Technician B is correct. A defective driver's blend door actuator could cause the driver to not be able to control the temperature on that side of the vehicle.

Answer C is incorrect. Only Technician B is correct.

Answer D is incorrect. Technician B is correct.

TASK B.1.4

41. The compressor clutch plate air gap is being inspected on a compressor. Technician A says that a feeler gauge can be used to make this measurement. Technician B says that the compressor clutch plate will slip when the compressor engages if this measurement is too wide. Who is correct?

A. A only
B. B only
C. Both A and B
D. Neither A nor B

Answer A is incorrect. Technician B is also correct.

Answer B is incorrect. Technician A is also correct.

Answer C is correct. Both Technicians are correct. A feeler gauge is the recommended tool to be used to check the compressor clutch plate air gap. If the air gap is too wide, the clutch plate could slip when the compressor engages.

Answer D is incorrect. Both Technicians are correct.

42. A vacuum actuator is being tested by applying 18 in. Hg (40.5 kPa absolute) of vacuum to the vacuum actuator. Technician A says the vacuum gauge reading should remain steady for at least one minute. Technician B says if the gauge reading drops slowly, the actuator is leaking. Who is correct?

TASK C.2.4

A. A only
B. B only
C. Both A and B
D. Neither A nor B

Answer A is incorrect. Technician B is also correct.

Answer B is incorrect. Technician A is also correct.

Answer C is correct. Both Technicians are correct. The vacuum level should hold for at least a minute when tested with a vacuum pump. If the vacuum level drops, the actuator is leaking and should be replaced.

Answer D is incorrect. Both Technicians are correct.

43. What function does the A/C clutch coil perform for the A/C system?

TASK C.1.3

A. Prevents the voltage spike from damaging other components
B. Creates magnetism to attract the clutch drive plate
C. Limits current flow to prevent the A/C fuse from overheating
D. Protects the A/C system by opening and closing

Answer A is incorrect. The compressor clutch clamping diode handles the voltage spike created by de-energizing the A/C clutch coil.

Answer B is correct. The A/C clutch coil creates a magnetic field when power and ground are connected to it. This magnetic field attracts the front clutch plate to be connected to the front pulley.

Answer C is incorrect. The coil does not limit current flow. It is a fixed coil that creates magnetism when energized.

Answer D is incorrect. The A/C clutch coil does not open and close. It is a fixed coil that creates magnetism when energized.

44. A cable-controlled heater control valve is being diagnosed. The control knob is very difficult to move and the heater valve does move slowly. Technician A says the cable housing clamp may be loose at the control head (panel) end. Technician B says the cable may be rusted inside the cable housing. Who is correct?

TASK C.2.3

A. A only
B. B only
C. Both A and B
D. Neither A nor B

Answer A is incorrect. A loose cable clamp would not cause the knob to be hard to turn and the valve would likely not move either.

Answer B is correct. Only Technician B is correct. Rust in the cable housing could cause the control knob to be difficult to move as well as the valve moving slowly.

Answer C is incorrect. Only Technician B is correct.

Answer D is incorrect. Technician B is correct.

TASK C.3.9, C.3.10

45. Technician A says that some computer-controlled A/C system actuator motors are automatically calibrated in the self-diagnostic mode. Technician B says that some computer-controlled A/C system actuator motors require a scan tool for calibration after replacement. Who is correct?

A. A only
B. B only
C. Both A and B
D. Neither A nor B

Answer A is incorrect. Technician B is also correct.

Answer B is incorrect. Technician A is also correct.

Answer C is correct. Both Technicians are correct. Electronic actuators can be calibrated on some vehicles by using the self-diagnostic mode on the control head. Some electronic actuators require a scan tool to calibrate the position after replacement.

Answer D is incorrect. Both Technicians are correct.

TASK A.5

46. Which of these is most likely to be performed during an A/C system recovery procedure?

A. Refrigerant is removed from the vehicle.
B. Refrigerant is added to the vehicle.
C. Most of the A/C system oil is removed the vehicle.
D. Refrigerant is tested for contaminants.

Answer A is correct. Refrigerant is removed from the A/C system during a recovery process. Most A/C recovery machines will filter and weigh the refrigerant being removed.

Answer B is incorrect. Refrigerant is added to the A/C system during the recharge process.

Answer C is incorrect. The A/C system oil can only be removed by flushing or replacing the A/C system's major components.

Answer D is incorrect. The refrigerant is tested for contaminants during the identifier test.

TASK A.14

47. A mode actuator is being diagnosed on a late-model vehicle. Technician A says that a test light could be used to measure the voltage supplied to the actuator. Technician B says that a digital storage oscilloscope (DSO) could be used to detect an intermittent electrical fault at the actuator. Who is correct?

A. A only
B. B only
C. Both A and B
D. Neither A nor B

Answer A is incorrect. A test light should never be used to measure voltage on an electronic circuit due to possibly damaging the circuit. Only high impedance test equipment such as a digital multi-meter should be used for this purpose.

Answer B is correct. Only Technician B is correct. A DSO is often used to detect intermittent faults in electronic circuits. These tools display a waveform of the voltage or current that is present in the circuit being tested.

Answer C is incorrect. Only Technician B is correct.

Answer D is incorrect. Technician B is correct.

48. The refrigerant is being recovered from an A/C system. Five minutes after the recovery process is complete, the low-side pressure remains in vacuum. This condition indicates:

TASK A.5

A. There is still some refrigerant in the system.
B. There is excessive oil in the refrigerant system.
C. The refrigerant system is completely recovered.
D. There is excessive moisture in the refrigerant system.

Answer A is incorrect. The low-side pressure would rise above vacuum if refrigerant remained in the system.

Answer B is incorrect. There is no data in this scenario that describes the oil volume of the system.

Answer C is correct. The recovery process is complete when the system remains in vacuum for a few minutes after the machine has shut off.

Answer D is incorrect. There is no data in this scenario that describes the moisture content of the system.

49. A refrigerant identifier is connected to an A/C system and gives the reading of 99 percent R134a and 1 percent air. Technician A says that this system can be safely recovered into the R134a recovery machine. Technician B says that this system may have had previous repairs due to the small percentage of air. Who is correct?

TASK A.5

A. A only
B. B only
C. Both A and B
D. Neither A nor B

Answer A is incorrect. Technician B is also correct.

Answer B is incorrect. Technician A is also correct.

Answer C is correct. Both Technicians are correct. Systems with 99 percent R134a and 1 percent air can be recovered. The recovery machine will remove small amounts of air during the filtering process. It is possible that the system has had a previous repair since it has some air in it.

Answer D is incorrect. Both Technicians are correct.

50. Technician A says that the duct temperature will vary depending on the ambient temperature. Technician B says that the system pressures should vary with changes in ambient temperature. Who is correct?

TASK A.2, A.3

A. A only
B. B only
C. Both A and B
D. Neither A nor B

Answer A is incorrect. Technician B is also correct.

Answer B is incorrect. Technician A is also correct.

Answer C is correct. Both Technicians are correct. Ambient temperature will affect both duct temperature as well as system pressures. Higher ambient temperatures will increase both duct temperature and system pressures. Duct temperature should typically be between 40 and 50°F while running a performance test.

Answer D is incorrect. Both Technicians are correct.

SECTION

7 Appendices

PREPARATION EXAM ANSWER SHEET FORMS

ANSWER SHEET

1. ____________
2. ____________
3. ____________
4. ____________
5. ____________
6. ____________
7. ____________
8. ____________
9. ____________
10. ____________
11. ____________
12. ____________
13. ____________
14. ____________
15. ____________
16. ____________
17. ____________
18. ____________
19. ____________
20. ____________
21. ____________
22. ____________
23. ____________
24. ____________
25. ____________
26. ____________
27. ____________
28. ____________
29. ____________
30. ____________
31. ____________
32. ____________
33. ____________
34. ____________
35. ____________
36. ____________
37. ____________
38. ____________
39. ____________
40. ____________
41. ____________
42. ____________
43. ____________
44. ____________
45. ____________
46. ____________
47. ____________
48. ____________
49. ____________
50. ____________

ANSWER SHEET

1. ______________
2. ______________
3. ______________
4. ______________
5. ______________
6. ______________
7. ______________
8. ______________
9. ______________
10. ______________
11. ______________
12. ______________
13. ______________
14. ______________
15. ______________
16. ______________
17. ______________
18. ______________
19. ______________
20. ______________
21. ______________
22. ______________
23. ______________
24. ______________
25. ______________
26. ______________
27. ______________
28. ______________
29. ______________
30. ______________
31. ______________
32. ______________
33. ______________
34. ______________
35. ______________
36. ______________
37. ______________
38. ______________
39. ______________
40. ______________
41. ______________
42. ______________
43. ______________
44. ______________
45. ______________
46. ______________
47. ______________
48. ______________
49. ______________
50. ______________

ANSWER SHEET

1. ____________
2. ____________
3. ____________
4. ____________
5. ____________
6. ____________
7. ____________
8. ____________
9. ____________
10. ____________
11. ____________
12. ____________
13. ____________
14. ____________
15. ____________
16. ____________
17. ____________
18. ____________
19. ____________
20. ____________
21. ____________
22. ____________
23. ____________
24. ____________
25. ____________
26. ____________
27. ____________
28. ____________
29. ____________
30. ____________
31. ____________
32. ____________
33. ____________
34. ____________
35. ____________
36. ____________
37. ____________
38. ____________
39. ____________
40. ____________
41. ____________
42. ____________
43. ____________
44. ____________
45. ____________
46. ____________
47. ____________
48. ____________
49. ____________
50. ____________

ANSWER SHEET

1. ____________
2. ____________
3. ____________
4. ____________
5. ____________
6. ____________
7. ____________
8. ____________
9. ____________
10. ____________
11. ____________
12. ____________
13. ____________
14. ____________
15. ____________
16. ____________
17. ____________
18. ____________
19. ____________
20. ____________
21. ____________
22. ____________
23. ____________
24. ____________
25. ____________
26. ____________
27. ____________
28. ____________
29. ____________
30. ____________
31. ____________
32. ____________
33. ____________
34. ____________
35. ____________
36. ____________
37. ____________
38. ____________
39. ____________
40. ____________
41. ____________
42. ____________
43. ____________
44. ____________
45. ____________
46. ____________
47. ____________
48. ____________
49. ____________
50. ____________

ANSWER SHEET

1. ______________
2. ______________
3. ______________
4. ______________
5. ______________
6. ______________
7. ______________
8. ______________
9. ______________
10. ______________
11. ______________
12. ______________
13. ______________
14. ______________
15. ______________
16. ______________
17. ______________
18. ______________
19. ______________
20. ______________
21. ______________
22. ______________
23. ______________
24. ______________
25. ______________
26. ______________
27. ______________
28. ______________
29. ______________
30. ______________
31. ______________
32. ______________
33. ______________
34. ______________
35. ______________
36. ______________
37. ______________
38. ______________
39. ______________
40. ______________
41. ______________
42. ______________
43. ______________
44. ______________
45. ______________
46. ______________
47. ______________
48. ______________
49. ______________
50. ______________

ANSWER SHEET

1. ______________
2. ______________
3. ______________
4. ______________
5. ______________
6. ______________
7. ______________
8. ______________
9. ______________
10. ______________
11. ______________
12. ______________
13. ______________
14. ______________
15. ______________
16. ______________
17. ______________
18. ______________
19. ______________
20. ______________
21. ______________
22. ______________
23. ______________
24. ______________
25. ______________
26. ______________
27. ______________
28. ______________
29. ______________
30. ______________
31. ______________
32. ______________
33. ______________
34. ______________
35. ______________
36. ______________
37. ______________
38. ______________
39. ______________
40. ______________
41. ______________
42. ______________
43. ______________
44. ______________
45. ______________
46. ______________
47. ______________
48. ______________
49. ______________
50. ______________

Glossary

Access Valve A term used for service port and service valve.

Accumulator A tank located between the evaporator and compressor to receive the refrigerant that leaves the evaporator, so constructed as to ensure that no liquid refrigerant will enter the compressor.

Actuator A device that transfers a vacuum or electric signal to a mechanical motion, typically performing an on/off or open/close function.

Adapter A device or fitting that permits different size parts or components to be fastened or connected to each other.

Aftermarket A term given to a device or accessory that is added to a vehicle after original manufacture, such as an air conditioning system.

Air Gap The space between two components, such as between the rotor and armature of a clutch.

Ambient Sensor A thermistor used in automatic temperature control units to sense ambient temperature.

Armature The part of the clutch that mounts onto the crank-shaft and engages with the rotor when energized.

Atmospheric Pressure Air pressure at a given altitude. At sea level, atmospheric pressure is 14.696 psia (101.329 kPa absolute).

Automatic Temperature Control A type of heating and air conditioning system that allows the driver to select a desired temperature on the control head. The A/C computer calculates the necessary output to adjust the cabin temperature.

Back Seat (service valve) Turning the valve stem to the left (ccw) as far as possible back seats the valve. The valve outlet to the system is open and the service port is closed.

Barb Fitting A fitting that slips inside a hose and is held in place with a gear-type clamp. Ridges (barbs) on the fitting prevent the hose from slipping off.

BCM An abbreviation for body control module.

Blower Relay An electrical device used to control the function or speed of a blower motor.

Boiling Point The temperature at which a liquid changes to a vapor.

Break a Vacuum The next step after evacuating a system. The vacuum should be broken with refrigerant or other suitable dry gas, not ambient air or oxygen.

By-Pass An alternate passage that may be used instead of the main passage.

By-Pass Hose A hose that is generally small and is used as an alternate passage to bypass a component or device.

CAA Abbreviation for the Clean Air Act.

Can Tap A device used to pierce, dispense, and seal small cans of refrigerant.

Can Tap Valve A valve found on a can tap that is used to control the flow of refrigerant.

Cap (1) A protective cover. (2) An abbreviation for capillary (tube) or capacitor.

Cap Tube A tube with a calibrated inside diameter and length used to control the flow of refrigerant, such as rgat, between the remote bulb to the expansion valve.

Celsius (C) A metric temperature scale using zero as the freezing point of water. The boiling point of water is 100°C (212°F).

Certified Having a certificate awarded or issued to those that have demonstrated appropriate competence through testing and/or practical experience.

CFC-12 A term used for Refrigerant-12.

Charge A specific amount of refrigerant or oil by volume or weight.

Check Valve A device that prevents refrigerant from flowing in the opposite direction when the unit is shut off.

Clean Air Act (CAA) A Title IV amendment signed into law in 1990 that established national policy relative to the reduction and elimination of ozone-depleting substances.

Clockwise (cw) A left to right rotation or motion.

Clutch An electro-mechanical device mounted on the air conditioning compressor used to start and stop compressor action, thereby controlling refrigerant circulating through the system.

Clutch Coil The electrical part of a clutch assembly. When electrical power is applied to the clutch coil, the clutch is engaged to start and stop compressor action.

Compound Gauge A gauge that registers both pressure and vacuum (above and below atmospheric pressure); used on the low side of the systems.

Compressor-Shaft Seal An assembly consisting of springs, snap rings, o-rings, shaft seal, seal sets, and gasket, mounted on the compressor crankshaft to permit the shaft to be turned without a loss of refrigerant or oil.

Contaminated A term used when referring to a refrigerant cylinder or a system that is known to contain foreign substances such as other incompatible or hazardous refrigerants.

Coolant Liquid that circulates in an engine cooling system.

Coolant Heater A component used to aid engine starting and reduce the wear caused by cold starting.

Coolant Hydrometer A tester designed to measure coolant-specific gravity and determine antifreeze protection.

Cooling System System for circulating coolant.

Counterclockwise (ccw) A direction, right to left, opposite of which a clock turns.

Cracked Position A mid-seated or open position.

Cycle Clutch Time (total) Time elapsed from the moment the clutch engages until it disengages, then reengages. Total time is equal to on-time plus off-time for one cycle.

Cycling Clutch Pressure Switch A pressure-actuated electrical switch used to cycle the compressor at a predetermined pressure.

Cycling Clutch System An air conditioning system in which the air temperature is controlled by starting and stopping the compressor with a thermostat or pressure control.

Department of Transportation (DOT) A federal agency charged with regulation and control of the shipment of all hazardous materials.

Depressing Pin A pin located in the end of a service hose to press (open) a Schrader-type valve.

Digital Multi-Meter (DMM) A device used in testing voltage, amperage, resistance and continuity of electrical circuits and components. This term is used more commonly than its equivalent term, digital volt ohmmeter (DVOM).

Digital Volt Ohmmeter (DVOM) A device used in testing voltage, amperage, resistance and continuity of electrical circuits and components. This term is used interchangeably with digital multi-meter (DMM); the latter term is more commonly used.

Disarm To turn off; to disable a device or circuit.

Dry Nitrogen The element nitrogen (N) that has been processed to ensure that it is free of moisture.

Dual System Two systems, usually refers to two evaporators in an air conditioning system, one in the front and one in the rear of the vehicle, driven off a single compressor and condenser system.

Duct A tube or passage used to provide a means to transfer air or liquid from one point or place to another.

EATC Abbreviation for electronic automatic temperature control.

ECC Abbreviation for electronic climate control.

Electronic Control Unit (ECU) A term used for a computer.

Environmental Protection Agency (EPA) An agency of the U.S. government that is charged with the responsibility of protecting the environment and enforcing the Clean Air Act (CAA) of 1990.

EPA Environmental Protection Agency.

Etch An intentional or unintentional erosion of a metal surface generally caused by an acid.

Evacuate To create a vacuum within a system to remove all traces of air and moisture.

Evaporator Core The tube and fin assembly located inside the evaporator housing. The refrigerant fluid picks up heat in the evaporator core when it changes into a vapor.

Expansion Tank An auxiliary tank that is usually connected to the inlet tank of a radiator and which provides additional storage space for heated coolant. Often called a coolant recovery tank.

External Snap Ring A snap ring found on the outside of a part, such as a shaft.

Fan Relay A relay for the cooling and/or auxiliary fan motors.

Fill Neck The part of the radiator on which the pressure cap is attached. Most radiators, however, are filled via the recovery tank.

Filter A device used with the dryer or as a separate unit to remove foreign material from the refrigerant.

Filter Dryer A device that has a filter to remove foreign material from the refrigerant and a desiccant to remove moisture from the refrigerant.

Flare A flange or cone-shaped end applied to a piece of tubing to provide a means of fastening to a fitting.

Forced Air Air that is moved mechanically, such as by a fan or blower.

Front Seat Closing off the line, leaving the compressor open to the service port fitting. This allows service to the compressor without purging the entire system. Never operate the system with the valves front seated.

Functional Test A term used for performance test.

Fusible Link A type of fuse made of a special wire that melts to open a circuit when current draw is excessive.

Gasket A thin layer of material or composition that is placed between two machined surfaces to provide a leakproof seal between them.

Gauge A tool of a known calibration used to measure components. For example, a feeler gauge is used to measure the air gap between a clutch rotor and armature.

Graduated Container A measure, such as a beaker or measuring cup, that has a graduated scale for the measure of a liquid.

Ground A general term given to the negative (–) side of an electrical system.

Grounded An intentional or unintentional connection of a wire, positive (+) or negative (–), to the ground. A short circuit is said to be grounded.

Gross Weight The weight of a substance or matter that includes the weight of its container.

HCFC Abbreviation for hydrochlorofluorocarbon refrigerant.

Header Tank The top and bottom tanks (downflow) or side tanks (crossflow) of a radiator. The tanks in which coolant is accumulated or received.

Heater Core A radiator-like heat exchanger located in the case/duct system through which coolant flows to provide heat to the vehicle interior.

Heat Exchanger An apparatus in which heat is transferred from one medium to another on the principle that heat moves to an object with less heat.

HI The designation for high as in blower speed or system mode.

High-Side Gauge The correct side gauge on the manifold used to read refrigerant pressure in the high side of the system.

High-Side Hand Valve The high-side valve on the manifold set used to control flow between the high side and service ports.

High-Side Service Valve A device located on the discharge side of the compressor; this valve permits the service technician to check the high-side pressures and perform other necessary operations.

High-Side Switch See Pressure Switch.

High-Torque Clutch A heavy-duty clutch assembly used on some vehicles known to operate with higher-than-average head pressure.

Hot Knife A knife-like tool that has a heated blade used for separating objects, such as evaporator cases.

Hub The central part of a wheel-like device, such as a clutch armature.

Hygiene A system of rules and principles intended to promote and preserve health.

Hygroscopic Readily absorbing and retaining moisture.

Idler A pulley device that keeps the belt whip out of the drive belt of an automotive air conditioner. The idler is used as a means of tightening the belt.

Idler Pulley A pulley used to tension or torque the belt(s).

Idle Speed The speed (RPM) at which the engine runs while at rest (idle).

In-Car Temperature Sensor A thermistor used in automatic temperature control units for sensing the in-car temperature. Also see Thermistor.

Insert Fitting A fitting that is designed to fit inside, such as a barb fitting that fits inside a hose.

Internal Snap Ring A snap ring used to hold a component or part inside a cavity or case.

Jumper A wire used to temporarily bypass a device or component for the purpose of testing.

Kilogram (kg) A unit of measure in the metric system. One kilogram is equal to 2.205 pounds in the English system.

KiloPascal (kPa) A unit of measure in the metric system. One kilopascal (kPa) is equal to 0.145 pound per square inch (psi) in the English system.

kPa An abbreviation for kiloPascal.

Liquid A state of matter; a column of fluid without solids or gas pockets.

Low-Refrigerant Switch A switch that senses low pressure due to a loss of refrigerant and stops compressor action. Some alert the operator and/or set a trouble code.

Low-Side Gauge The left-side gauge on the manifold used to read refrigerant pressure in the low side of the system.

Low-Side Hand Valve The manifold valve used to control flow between the low side and service ports of the manifold.

Low-Side Service Valve A device located on the suction side of the compressor that allows the service technician to check low-side pressures and perform other necessary service operations.

Manifold A device equipped with a hand shutoff valve. Gauges are connected to the manifold for use in system testing and servicing.

Manifold and Gauge Set A manifold complete with gauges and charging hoses.

Manifold Hand Valve Valves used to open and close passages through the manifold set.

Manufacturer's Procedures Specific step-by-step instructions provided by the manufacturer for the assembly, disassembly, installation, replacement, and/or repair of a particular product manufactured by them.

MAX A mode, maximum, for heating or cooling. Selecting MAX generally overrides all other conditions that may have been programmed.

Mid-Positioned The position of a stem-type service valve where all fluid passages are interconnected. Also referred to as "cracked."

Motor An electrical device that produces a continuous turning motion. A motor is used to propel a fan blade or a blower wheel.

MSDS Abbreviation for material safety data sheet.

Net Weight The weight of a product only; container and packaging not included.

Neutral On neither side of positive or negative; the position of gears when force is not being transmitted.

Noncycling Clutch An electro-mechanical compressor clutch that does not cycle on and off as a means of temperature control. It is used to turn the system on when cooling is desired and off when cooling is not desired.

OEM Abbreviation for original equipment manufacturer.

Off-the-Road Generally refers to vehicles that are not licensed for road use, such as harvesters, bulldozers, and so on.

Ohmmeter An electrical instrument used to measure the resistance in ohms of a circuit or component.

Open Not closed. An open switch, for example, breaks an electrical circuit.

Orifice A calibrated opening in a tube or pipe to regulate the flow of a fluid or liquid.

O-ring A synthetic rubber or plastic gasket with a round- or square-shaped cross section.

OSHA Abbreviation for the Occupational Safety and Health Administration.

Outside Temperature Sensor A term used for ambient sensor.

Overcharge Indicates that too much refrigerant or refrigeration oil is added to the system.

Overload Anything in excess of the design criteria. An overload will generally cause the protective device, such as a fuse or pressure relief, to open.

Ozone Friendly Any product that does not pose a hazard or danger to the ozone.

PAG Abbreviation for polyalkylene glycol.

Park Generally refers to a component or mechanism that is at rest.

PCM Abbreviation for power train control module.

Performance Test Readings of the temperature and pressure under controlled conditions to determine if an air conditioning system is operating at full efficiency.

Piercing Pin The part of a saddle valve that is used to pierce a hole in the tubing.

Pin-Type Connector A single or multiple electrical connector that is round- or pin-shaped and fits inside a matching connector.

Poly Belt A term used for serpentine belt.

Polyalkylene Glycol (PAG) A synthetic-based lubricant that is used in R134a air conditioning systems.

Polyester (ESTER) A synthetic oil-like lubricant that is occasionally recommended for use in an HFC-134a system. This lubricant is compatible with both HFC-134a and CFC-12.

Positive Pressure Any pressure above atmospheric.

Pound A weight measure 16 ounces.

Pound of Refrigerant A term often used by technicians when referring to a small can of refrigerant, although it actually contains less than 16 ounces.

Power Module Controls the operation of the blower motor in an automatic temperature control system.

Predetermined A set of fixed values or parameters that have been programmed or otherwise fixed into an operating system.

Pressure Gauge A calibrated instrument for measuring pressure.

Pressure Sensor A variable sensor used to detect A/C pressure in the high side of the system.

Pressure Switch An electrical switch that is activated by a predetermined low or high pressure. A high-pressure switch is generally used for system protection; a low-pressure switch may be used for temperature control or system protection.

Propane A flammable gas used as a propellant for the halide leak detector.

PSIG Abbreviation for pounds per square inch gauge.

Purge To remove moisture and/or air from a system or a component by flushing with a dry gas such as nitrogen (N) to remove all refrigerant from the system.

Purity Test A static test that may be performed to compare the suspect refrigerant pressure to an appropriate temperature chart to determine its purity.

Radiation The transfer of heat without heating the medium through which it is transmitted.

Ram Air Air that is forced through the radiator and condenser coils by the movement of the vehicle or the action of the fan.

Receiver/Drier A tank-like vessel having a desiccant and used for the storage of refrigerant.

RECIR An abbreviation for the recirculate mode, as with air.

Recovery System A term often used to refer to the circuit inside the recovery unit used to recycle and/or transfer refrigerant from the air conditioning system to the recovery cylinder.

Recovery Tank An auxiliary tank, usually connected to the inlet tank of a radiator, which provides additional storage space for heated coolant.

Refrigerant-12 The refrigerant used in automotive air conditioners, as well as other air conditioning and refrigeration systems.

Relay An electrical switch device that is activated by a low-current source and controls a high-current device.

Reserve Tank A storage vessel for excess fluid. See Recovery Tank, Receiver/Drier, and Accumulator.

Resistor A voltage-dropping device that is usually wire-wound and provides a means of controlling fan speeds.

Respirator A mask or face shield worn in a hazardous environment to provide clean fresh air and/or oxygen.

Restrictor An insert fitting or device used to control the flow of refrigerant or refrigeration oil.

Retrofitting The name given to the procedure for converting R-12 A/C systems to be able to use R-134a refrigerant.

Rotor The rotating or freewheeling portion of a clutch; the belt slides on the rotor.

RPM Abbreviation for revolutions per minute; also, rpm or r/min.

Running Design Change A design change made during a current model/year production.

Saddle Valve A two-part accessory valve that may be clamped around the metal part of a system hose to provide access to the air conditioning system for service.

SAE Abbreviation for Society of Automotive Engineers.

Schrader Valve A spring-loaded valve similar to a tire valve. The Schrader valve is located inside the service valve fitting and is used on some control devices to hold refrigerant in the system. Special adapters must be used with the gauge hose to allow access to the system.

Seal Generally refers to a compressor shaft oil seal; matching shaft-mounted seal face and front head-mounted seal seat to prevent refrigerant and/or oil from escaping. May also refer to any gasket or o-ring used between two mating surfaces for the same purpose.

Seal Seat The part of a compressor shaft seal assembly that is stationary and matches the rotating part, known as the seal face or shaft seal.

Serpentine Belt A flat or V-groove belt that winds through all of the engine accessories to drive them off the crankshaft pulley.

Service Port A fitting found on the service valves and some control devices; the manifold set hoses are connected to this fitting.

Service Procedure A suggested routine for the step-by-step act of troubleshooting, diagnosing and/or repairs.

Service Valve See High-Side (Low-Side) service valve.

Shaft Key A soft metal key that secures a member on a shaft to prevent it from slipping.

Shaft Seal See Compressor-Shaft Seal.

Short of Brief Duration For example, short cycling. Also refers to an intentional or unintentional grounding of an electrical circuit.

Shut-Off Valve A valve that provides positive shut-off of a fluid or vapor passage.

Snap Ring A metal ring used to secure and retain a component to another component.

Society of Automotive Engineers A professional organization of the automotive industry. Founded in 1905 as the Society of Automobile Engineers, the SAE is dedicated to providing technical information and standards to the automotive industry.

Solenoid Valve An electromagnetic valve controlled remotely by electrically energizing and de-energizing a coil.

Solid State Referring to electronics, consisting of semiconductor devices and other related nonmechanical components.

Spade-Type Connector A single or multiple electrical connector that has flat, spade-like mating provisions.

Specifications Design characteristics of a component or assembly noted by the manufacturer. Specifications for a vehicle include fluid capacities, weights, and other pertinent maintenance information.

Spike In our application, an electrical spike. An unwanted momentary high-energy electrical surge.

Spring Lock Fitting A special fitting using a spring to lock the mating parts together forming a leak-proof joint.

Squirrel-Cage Blower A blower wheel designed to provide a large volume of air with a minimum of noise. The blower is more compact than the fan and air can be directed more efficiently.

Stabilize To make steady.

Subsystem A system within a system.

Sun Load Heat intensity and/or light intensity produced by the sun.

Superheat Switch An electrical switch activated by an abnormal temperature-pressure condition (a superheated vapor); used for system protection.

Tank A term used for header tank and expansion tank.

Tare Weight The weight of the packaging material.

Temperature Door A door within the case/duct stem to direct air through the heater and/or evaporator core.

Temperature Switch A switch actuated by a change in temperature at a predetermined point.

Tension Gauge A tool for measuring the tension of a belt.

Thermistor A temperature-sensing resistor that has the ability to change values with changing temperature.

Torque A turning force, for example, the force required to seal a connection; measured in (English) foot-pounds (ft-lb) or inch-pounds (in-lb); (metric) Newton-meters (N·m).

Triple Evacuation A process of evacuation that involves three pumpdowns and two system purges with an inert gas, such as dry nitrogen (N).

TXV Abbreviation for thermostatic expansion valve.

Ultraviolet (UV) The part of the electromagnetic spectrum emitted by the sun that lies between visible violet light and X-rays.

Vacuum Gauge A gauge used to measure below atmospheric pressure.

Vacuum Motor A device designed to provide mechanical control by the use of a vacuum.

Vacuum Pump A mechanical device used to evacuate the refrigeration system to rid it of excess moisture and air.

Vacuum Signal The presence of a vacuum.

V-Belt A rubber-like continuous loop placed between the engine crankshaft pulley and accessories to transfer rotary motion of the crankshaft to the accessories.

Ventilation The act of supplying fresh air to an enclosed space, such as the inside of an automobile.

V-Groove Belt A term used for V-belt.

Voltmeter A device used to measure volt(s).

Wiring Harness A group of wires wrapped in a shroud for the distribution of power from one point to another point.

Notes

Notes

Notes

Notes

Notes